BIONANOTECHNOLOGY

BIONANOTECHNOLOGY

Lessons from Nature

David S. Goodsell, Ph.D.
Department of Molecular Biology
The Scripps Research Institute
La Jolla, California

 **WILEY-LISS**

A JOHN WILEY & SONS, INC., PUBLICATION

For general information on our other products and services please contact our Customer Care Department within the U.S. at 877-762-2974, outside the U.S. at 317-572-3993 or fax 317-572-4002.

Wiley also publishes its books in a variety of electronic formats. Some content that appears in print, however, may not be available in electronic format.

Library of Congress Cataloging-in-Publication Data:

Goodsell, David S.
    Biotechnology : lessons from nature / David S. Goodsell.
        p. ; cm.
Includes bibliographical references and index.
        ISBN 0-471-41719-X (cloth : alk. paper)
    1. Biomolecules.   2. Nanotechnology.   3. Biotechnology.
    [DNLM: 1. Biotechnology.   2. Nanotechnology.   QT 36 G655b 2004]   I.
Title.
    QP514.2.G658 2004
    660.6—dc21   2003006943

Printed in the United States of America

10 9 8 7 6 5 4 3 2

CONTENTS

PREFACE

Today is the most exciting time to be working in nanotechnology, and bionanotechnology in particular. Chemistry, biology, and physics have revealed an immense amount of information on molecular structure and function, and now we are poised to make use of it for atomic-level engineering. New discoveries are being made every day, and clever people are pressing these discoveries into service in every imaginable (and unimaginable) way.

In this book, I present many of the lessons that may be learned from biology and how they are being applied to nanotechnology. The book is divided into three basic parts. In the first part, I explore the properties of the nanomachines that are available in cells. In Chapter 2, I present the unfamiliar world of bionanomachines and go on a short tour of the natural nanomachinery that is available for our use. Chapter 3 provides an overview of the techniques that are available in biotechnology for harnessing and modifying these nanomachines.

In the second part, I look to these natural nanomachines for guidance in the building of our own nanomachinery. By surveying what is known about biological molecules, we can isolate the general principles of structure and function that are used to construct functional nanomachines. These include general structural principles, presented in Chapter 4, and functional principles, described in Chapter 5.

The book finishes with two chapters on applications. Chapter 6 surveys some of the exciting applications of bionanotechnology that are currently under study. The final chapter looks to the future, speculating about what we might expect.

Bionanotechnology is a rapidly evolving field, which encompasses a diverse collection of disciplines. This book necessarily omits entire sectors of research and interest and is unavoidably biased by my own interests and my own background as a structural biologist. Biomolecular science still holds many deep mysteries and exciting avenues for study, which should provide even more source material for bionanotechnology in the coming decades. I invite you to explore the growing literature in this field, using this book as an invitation for further reading.

I thank Arthur J. Olson for many useful discussions during the writing of this book.

DAVID S. GOODSELL

THE QUEST FOR NANOTECHNOLOGY

The principles of physics, as far as I can see, do not speak against the possibility of maneuvering things atom by atom. It is not an attempt to violate any laws; it is something, in principle, that can be done; but in practice, it has not been done because we are too big.

—Richard Feynman*

Nanotechnology is available, today, to anyone with a laboratory and imagination. You can create custom nanomachines with commercially available kits and reagents. You can design and build nanoscale assemblers that synthesize interesting molecules. You can construct tiny machines that seek out cancer cells and kill them. You can build molecule-size sensors for detecting light, acidity, or trace amounts of poisonous metals. Nanotechnology is a reality today, and nanotechnology is accessible with remarkably modest resources.

What is nanotechnology? Nanotechnology is the ability to build and shape matter one atom at a time. The idea of nanotechnology was first presented by physicist Richard Feynman. In a lecture entitled "Room at the Bottom," he unveiled the possibilities available in the molecular world. Because ordinary matter is built of so many atoms, he showed that there is a

---

*All opening quotes are taken from Richard P. Feynman's 1959 talk at the California Institute of Technology, as published in the February 1960 issue of CalTech's *Engineering and Science.*

Bionanotechnology: Lessons from Nature. By David S. Goodsell
ISBN 0-471-41719-X Copyright © 2004 John Wiley & Sons, Inc.

remarkable amount of space within which to build. Feynman's vision spawned the discipline of nanotechnology, and we are now amassing the tools to make his dream a reality.

But atoms are almost unbelievably small; a million times smaller than objects in our familiar world. Their properties are utterly foreign, so our natural intuition and knowledge of the meter-scale world is useless at best and misleading at worst. How can we approach the problem of engineering at the atomic scale?

When men and women first restructured matter to fit their needs, an approach opposite from nanotechnology was taken. Instead of building an object from the bottom up, atom-by-atom, early craftsmen invented a top-down approach. They used tools to shape and transform existing matter. Clay, plant fibers, and metals were shaped, pounded, and carved into vessels, clothing, and weapons. With some added sophistication, this approach still accounts for the bulk of all products created by mankind. We still take raw materials from the earth and physically shape them into functional products.

Mankind did not make any concerted effort to shape the atoms in manufactured products until medieval times, when alchemists sowed the seeds of the modern science of chemistry. During their search for the secrets of immortality and the transmutation of lead to gold, they developed methods for the willful combination of atoms. Chemical reaction, purification, and characterization are all tools of the alchemists. Today, chemists build molecules of defined shape and specified properties. Chemical reactions are understood, and tailored, at the atomic level. Most of chemistry, however, is performed at a bulk level. Large quantities of pure materials are mixed and reacted, and the desired product is purified from the mixture of molecules that are formed. Nonetheless, chemistry is nanotechnology—the willful combination of atoms to form a desired molecule. But it is nanotechnology on a bulk scale, controlled by statistical mechanics rather than controlled atom-by-atom at the nanometer scale.

We are now in the midst of the second major revolution of nanotechnology. Now, scientists are attempting modify matter one atom at a time.

Some envision a nanotechnology closely modeled after our own macroscopic technology. This new field has been dubbed *molecular nanotechnology* for its focus on creating molecules individually atom-by-atom. K. Eric Drexler has proposed methods of constructing molecules by forcibly pressing atoms together into the desired molecular shapes, in a process dubbed "mechanosynthesis" for its parallels with macroscopic machinery and engineering. With simple raw materials, he envisions building objects in an assembly-line manner by directly bonding individual atoms. The idea is compelling. The engineer retains direct control over the synthesis, through a physical connection between the atomic realm and our macroscopic world.

Central to the idea of mechanosynthesis is the construction of an *assembler*. This is a nanometer-scale machine that assembles objects atom-by-atom according to defined instructions. Nanotechnology aficionados have speculated that the creation of just a single working assembler would lead immediately to the "Two-Week Revolution." They tell us that as soon as a single assembler is built, all of the dreams of nanotechnology would be realized within days. Researchers could immediately direct this first assembler to build additional new assemblers. These assemblers would immediately allow construction of large-scale factories, filled with level upon level of assemblers for building macroscale objects. Nanotechnology would explode to fill every need and utterly change our way of life. Unfortunately, assemblers based on mechanosynthesis currently remain only an evocative idea.

The subject of this book is another approach to nanotechnology, which is available today to anyone with a moderately equipped laboratory. This is *bionanotechnology*, nanotechnology that looks to nature for its start. Modern cells build thousands of working nanomachines, which may be harnessed and modified to perform our own custom nanotechnological tasks. Modern cells provide us with an elaborate, efficient set of molecular machines that restructure matter atom-by-atom, exactly to our specifications. And with the well-tested techniques of biotechnology, we can extend the function of these machines for our own goals, modifying existing biomolecular nanomachines or designing entirely new ones.

BIOTECHNOLOGY AND THE TWO-WEEK REVOLUTION

The Two-Week Revolution has already occurred, although it has lasted for decades instead of weeks. Biotechnology uses the ready-made assemblers available in living cells to build thousands of custom-designed molecules to atomic specifications, including the construction of new assemblers. This has lead to myriad applications, including commercial production of hormones and drugs, elegant methods for diagnosing and curing infectious and genetic diseases, and engineering of organisms for specialized tasks such as bioremediation and disease resistance.

Biotechnology took several decades to gather momentum. The primary impediment has been the lack of basic knowledge of biomolecular processes and mechanisms. We have been given an incredible toolbox of molecular machinery, and we are only now beginning to learn how to use it. The key enabling technology, recombinant DNA, made the natural protein assembler of the cell available for use. The subsequent years have yielded numerous refinements on the technology, and numerous ideas on how it might be exploited.

Biotechnology has grown, and is still growing, with each new discovery in molecular biology. Further research into viral biology has led to improved vectors for delivering new genetic material. An explosion of enzymes for clipping, editing, ligating, and copying DNA, as well as efficient techniques for the chemical synthesis of DNA, has allowed the creation of complicated new genetic constructs. Engineered bacteria now create large quantities of natural proteins for medicinal use, mutated proteins for research, hybrid chimeric proteins for specialized applications, and entirely new proteins, if a researcher is bold enough to design a protein from scratch.

FROM BIOTECHNOLOGY TO BIONANOTECHNOLOGY

We are now poised to extend biotechnology into bionanotechnology. What is bionanotechnology, and how is it different from biotechnology? The two terms currently share an overlapped field of topics. I will define bionanotechnology here as applications that require human design and construction at the nanoscale level and will label projects as biotechnology when nanoscale understanding and design are not necessary. Biotechnology grew from the use of natural enzymes to manipulate the genetic code, which was then used to modify entire organisms. The atomic details were not really important—existing functionalities were combined to achieve the end goal. Today, we have the ability to work at a much finer level with a more detailed level of understanding and control. We have the tools to create biological machines atom-by-atom according to our own plans. Now, we must flex our imagination and venture into the unknown.

Bionanotechnology has many different faces, but all share a central concept: the ability to design molecular machinery to atomic specifications. Today, individual bionanomachines are being designed and created to perform specific nanoscale tasks, such as the targeting of a cancer cell or the solution of a simple computational task. Many are toy problems, designed to test our understanding and control of these tiny machines. As bionanotechnology matures, we will redesign the biomolecular machinery of the cell to perform large-scale tasks for human health and technology. Macroscopic structures will be built to atomic precision with existing biomolecular assemblers or by using biological models for assembly. Looking to cells, we can find atomically precise molecule-sized motors, girders, random-access memory, sensors, and a host of other useful mechanisms, all ready to be harnessed by bionanotechnology. And the technology for designing and constructing these machines in bulk scale is well worked out and ready for application today.

Nanomedicine will be the biggest winner. Bionanomachines work best in the environment of a living cell and so are tailored for medical applications. Complex molecules that seek out diseased or cancerous cells are already a reality. Sensors for diagnosing diseased states are under development. Replacement therapy, with custom-constructed molecules, is used today to treat diabetes and growth hormone deficiencies, with many other applications on the horizon.

Biomaterials are another major application of bionanotechnology. We already use biomaterials extensively. Look around the room and notice how much wood is used to build your shelter and furnishing and how much cotton, wool, and other natural fibers are used in your clothing and books. Biomaterials address our growing ecological sensitivity—biomaterials are strong but biodegradable. Biomaterials also integrate perfectly with living tissue, so they are ideal for medical applications.

The production of hybrid machines, part biological and part inorganic, is another active area of research in bionanotechnology that promises to yield great fruits. Bionanomachines, such as light sensors or antibodies, are readily combined with silicon devices created by microlithography. These hybrids provide a link between the nanoscale world of bionanomachines and the macroscale world of computers, allowing direct sensing and control of nanoscale events.

Finally, Drexler and others have seen biological molecules as an avenue to reach their own goal of mechanosynthesis using nanorobots. Certainly, biology provides the tools for building objects one atom at a time. Perhaps as our understanding grows, bionanomachines will be coaxed into building objects that are completely foreign to the biological blueprint.

This book explores these bionanomachines: their properties, their design principles, and the way they have been harnessed for our own applications. An exponentially growing body of information is being amassed, revealing the structure and function of individual biomolecules and their interactions within living cells. This information is a key resource for understanding the basic principles of nanomachinery: its structure, its function, and its integration into any larger application of nanotechnology. These existing, working nanomachines provide important lessons for the construction of our own nanotechnology, whether based directly on biology or constructed completely from our own imagination.

WHAT IS BIONANOTECHNOLOGY?

Nanotechnology and bionanotechnology are entirely new concepts, invented late in the twentieth century, and biotechnology has only been around for a few decades, so the scope of these fields is still being defined. With so many clever researchers working on all aspects of nanoscale structure, construction, and function, new examples that cross existing conceptual boundaries are appearing daily. Before getting started, it is worth spending a moment to compare the many technologies working at small scales and try to define the current scope of bionanotechnology.

Chemistry was the first science to manipulate molecules, starting when the first human beings cooked their food. Today, chemists design molecules and perform extensive, controlled syntheses to create them. Chemistry dif-

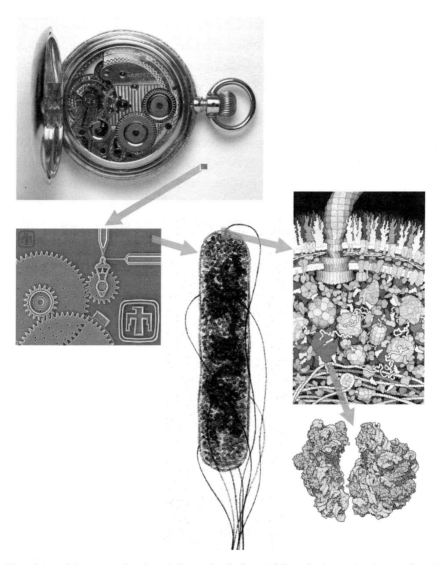

Figure 1-1 How big is bionanotechnology? Since the Industrial Revolution, scientists and engineers have constructed machines at an ever-smaller scale. Machines in our familiar world have moving parts in the range of millimeters to meters. As machining capabilities improved, tiny machines, such as the movement of a fine watch, extended the precision of machining to a fraction of a millimeter. Computer technology, with its ever-present pressure to miniaturize in order to improve performance, has driven the construction of tiny structures to even smaller ranges, with micrometer-scale construction of electronic components and tiny machines, like these tiny gears, created at the Sandia National Laboratories. Bionanotechnology operates at the smallest level, with machines in the range of 10 nm in dimension. The bacterium shown builds thousands of different bionanomachines, including a working nanoscale assembler, the ribosome, shown at lower right. Because these bionanomachines are composed of a finite, defined number of atoms, they represent a limit to the possible miniaturization of machinery. [MEMS gear photomicrograph from http://mems.sandia. gov/scripts/images.asp].

fers from bionanotechnology because it does not work at the level of individual molecules. There is no localization at the atomic level and no ability to address individual molecules. As a consequence of the bulk nature of chemistry, the molecules produced are generally limited to under a hundred atoms or so—syntheses of larger molecules are plagued by too many side reactions that form competing impurities.

Photolithography is widely used for the creation of computer hardware, and the growing field of MEMS is applying these technologies to the creation of microscale machines. Our entire information and communication technology relies on these methods. It relies on photographic techniques for reduction of scale and random deposition of atoms within the mask. Thus it is a macroscale technique scaled down to its finest limits.

Biotechnology harnesses biological processes and uses them for our own applications. In this book, I will limit the scope of biotechnology to applications that do not require atomic specification of biomolecules. For instance, researchers routinely use purified enzymes to cut and paste genetic instructions and add these back into cells. Knowledge of the atomic details are unimportant, just as knowledge of the type of ink used to print this page is not important for understanding of the words printed here.

Nanotechnology has been defined as engineering and manufacturing at nanometer scales, with atomic precision. The theoretical constructions popularized by K. Eric Drexler and the Foresight Institute are perhaps the most visible examples, and these are often further classified as "molecular nanotechnology." The positioning of individual argon atoms on a crystal surface by researchers at IBM is a successful example of nanotechnology.

Bionanotechnology is a subset of nanotechnology: atom-level engineering and manufacturing using biological precedents for guidance. It is also closely married to biotechnology but adds the ability to design and modify the atomic-level details of the objects created. Bionanomachines are designed to atomic specifications, they perform a well-defined three-dimensional molecular task, and, in the best applications, they contain mechanisms for individual control embedded in their structure.

BIONANOMACHINES IN ACTION

2

*I am inspired by the biological phenomena in which
chemical forces are used in repetitious fashion to produce
all kinds of weird effects (one of which is the author).*
                                          —Richard Feynman

As you read these words, 10,000 different nanomachines are at work inside your body. These are true nanomachines. Each one is a machine built to nanoscale specifications, with each atom precisely placed and connected to its neighbors. Your body is arguably the most complex mechanism in the known universe, and most of the action occurs at the nanoscale level. These nanomachines work in concert to orchestrate the many processes of life— eating and breathing, growing and repairing, sensing danger and responding to it, and reproducing.

Remarkably, many of these nanomachines will still perform their atom-sized functions after they are isolated and purified, provided that the environment is not too harsh. They do not have to be sequestered safely inside cells. Each one is a self-sufficient molecular machine. Already, these nanomachines have been pressed into service. Natural digestive enzymes like pepsin and lysozyme are so tough that they can be added to laundry detergent to help digest away stains. Amylases are used on an industrial scale to convert powdery starch into sweet corn syrup. The entire field of genetic engineering and biotechnology is made possible by a collection of DNA-

Bionanotechnology: Lessons from Nature. By David S. Goodsell
ISBN 0-471-41719-X   Copyright © 2004 John Wiley & Sons, Inc.

manipulating nanomachines, now available commercially. In general, natural bionanomachines are remarkably robust.

This chapter explores the bionanomachines made by living cells. They are different from the machines in our familiar world in many ways. They have been developed by the process of evolution (instead of intelligent design), which places unfamiliar restrictions on the process of design and the form of the final machine. Bionanomachines are also selected to perform their tasks in a very specific environment and are subject to the unfamiliar forces imposed by this environment. We must keep these differences in mind when trying to understand natural biomolecules, and we must keep these differences in mind when we use these natural bionanomachines as the starting point for our own bionanotechnology.

THE UNFAMILIAR WORLD OF BIONANOMACHINES

Biological machinery is different from anything we build with our familiar, human-sized technology. Natural biomolecules have organic, visceral, and often unbelievable shapes, unlike the tidy designs of toasters and tractors. They perform their jobs in a foreign environment, where jittery thermal motion is constantly pushing and pulling on their component parts. They are held together by a complex collection of bonding and nonbonding forces. At their small scale, bionanomachines are almost immune to the laws of gravity and inertia that dominate our machines. The world of bionanotechnology is an unfamiliar, shifting world that plays by different rules.

Gravity and Inertia are Negligible at the Nanoscale

Macroscopic objects, like bicycles and bridges, are dominated by the properties of mass. For centimeter-sized and meter-sized objects, physical properties such as friction, tensile strength, adhesion, and shear strength are comparable in magnitude to the forces imposed by inertia and gravity. So we can design picture hooks that are strong enough to hold up pictures and tires that will not fly apart when rotated at rapid speed. This balance changes, however, when we move to larger or smaller objects. As we move to larger objects, scaling laws shift the balance. Mass increases with the cube of the size of an object, and properties such as strength and friction increase linearly or with the square of size. The increase in inertia or weight can quickly overcome the increase in strength in a large structure such as a building. These scaling laws are quite familiar, and it is common sense to add extra support as we build larger and larger structures. We do not expect to be able to build a skyscraper a mile tall.

These scaling laws also apply in the opposite direction, with the opposite effect as we go to smaller and smaller machines. Micrometer-sized objects, like individual grains of sand or individual cells, already interact differently from macroscopic objects. Inertia is no longer a relevant property, so our intuition may lead to inappropriate conclusions. For example, E. M. Purcell described the surprising properties of bacterial cells swimming in water. These cells use a long corkscrew-shaped flagellum to propel themselves through the water. When the cell stops turning the flagellum, we might expect that the cell would slowly coast to a stop, like a submarine does in the ocean. However, because of the inertia scales differently relative to the viscous forces within the surrounding water, the cell actually stops in less than the diameter of an atom.

Gravity is also a negligible force when dealing with small objects. The actions of small objects are dominated by their interaction with neighboring objects. The molecules in water and air are in constant motion, continually battering small objects from all sides. So, fine dust stays suspended in the air instead of dropping quickly to the floor, and objects in water, if you look at them with a microscope, undergo random Brownian motion. The attractive forces between small objects are also stronger than the force of gravity. Flies take advantage of these attractive forces and can crawl up walls. Similarly, water droplets can hang from the ceiling because of these attractive forces.

Nanomachines Show Atomic Granularity

Nanoscale objects are built of discrete combinations of atoms that interact through specific atom-atom interactions. We cannot design nanomachines in a smoothly graded range of sizes. They must be composed of an integral number of atoms. For instance, we cannot design a nanoscale rotary motor like a macroscale motor, with a smooth ring surrounding an axle undergoing a smooth rotary motion. Instead, existing nanoscale rotary motors, such as ATP synthase or the bacterial flagellar motor, adopt several discrete rotary states that cycle one after the other (described in Chapter 5). There is not a smooth transition from one state to the next. Instead, the motor jumps from state to state when the appropriate chemical energy is applied. (Note that although smooth atom-scale motion is not observed in natural systems, theoretical nanoscale versions of axles and bearings have been proposed in molecular nanotechnology that take advantage of a mismatch in the number of atoms to smooth out atomic granularity.)

Because of atomic granularity, the typical continuous representations used in engineering are not appropriate. Bulk properties such as viscosity and friction are not defined for discrete atomic ensembles. Instead, individual atomic properties must be used. Quantum mechanics provides a deep understanding of the properties of atoms within biomolecules, but, fortunately, most of the basic properties may be understood qualitatively, through the use of a set of simplified rules. The central concept is the existence of covalent bonds, which connect atoms into stable molecules of defined geometry. Addition of a few rules to describe the interaction of atoms that are not bonded together—steric repulsion of nonbonded atoms, electrostatic interactions, and hydrogen bonds—allows understanding of most aspects of biomolecular structure and interaction. In general, biomolecules may be thought of as articulated chains of atoms that interact in a few well-defined ways. These qualitative rules are described in more detail in Chapter 4.

Thermal Motion is a Significant Force at the Nanoscale

Molecular nanotechnology seeks to create a "machine-phase" environment, with individual nanomachines organized like clockwork to form microscale and macroscale objects. Natural bionanomachinery takes a different approach, creating atomically precise nanomachinery but then enclosing them in a cellular space. The individual parts then interact through random motion and diffusion. In specialized applications machine-phase bionanostructures are used (two examples are presented in Chapter 5), but the bulk of the work done in cells is performed in the context of random, diffusive motion.

Bionanomachines operate in a chaotic environment. They are bombarded continually by water molecules. They will scatter randomly if not firmly held in place. Bionanomachines operate by forming interactions with other bionanomachines, fitting together and breaking apart in the course of action. If two molecules fit closely together and have the appropriate matching of chemical groups, they will interact over long periods of time. If the interactions are weaker, they will form only a temporary interaction before moving on to the next. By careful design of the strength of these interactions, bionanomachines can form stable molecular girders that last for years or delicate biosensors that fleetingly sense trace amounts of a molecule.

Cells are complex, with millions of individual proteins, and you might wonder whether diffusive motion is sufficient to allow interaction between the proper partners amidst all the competition. At the scale of the cell, diffusive motion is remarkably fast, so once again our intuition may play us false. If you release a typical protein inside a bacterial cell, within one-hundredth of a second, it is equally likely to be found anywhere in the cell. Place two molecules on opposite sides of the cell, and they are likely to interact within one second. As articulated by Hess and Mikhailov: "This result is remarkable: It tells us that *any* two molecules within a micrometer-size cell meet each other every second."

Bionanomachines Require a Water Environment

The form and function of biomolecules is dominated by two things: the chemistry of their component atoms and the unusual properties of the water surrounding them. The energetics of this interaction are quite different from anything we experience in our macroscopic world.

Water is an unusual substance, with specific preferences. Water molecules interact strongly with one another through hydrogen bonds. They do not lightly separate and interact with other molecules, unless these other molecules have something to offer. In biomolecules, regions that carry electronic charges and regions that are rich in nitrogen and oxygen atoms interact favorably with water. These regions easily dissolve into water solution.

Regions that are rich in carbon, however, cannot form the requisite hydrogen bonds and tend to be forced together in oily drops, minimizing contact with the surrounding water. This process has been termed the "hydrophobic effect," with the term hydrophobic referring to the "water-fearing" carbon atoms that avoid contact with water. Perhaps a better image is to think of water as an exclusive social clique that has no interest in carbon-rich conversation. The hydrophobic effect is described in more detail in Chapter 4.

The hydrophobic effect strongly shapes the form and function of a biological molecule. The geometry of the molecular chain alone allows a large range of conformations to be formed. If this were the whole story, life would not be possible—chains would only rarely form a single, defined structure. But when placed in water, biomolecules respond to the environment by folding into a conformation with the hydrophobic regions tucked away inside and the surface decorated with more water-loving groups. For proteins, the chain is most often forced into a compact globule. For DNA, the base pairs are sequestered safely inside, leaving the strongly charged phosphates on the surface. For lipids, many individual molecules are forced together to form membranes, with their hydrophobic atoms sandwiched between sheets of water-loving charged atoms. If designed carefully (as are all natural biological molecules), only a single structure is formed, creating a nanoscale machine with exactly the proper conformation to perform its duty (Figure 2-1).

MODERN BIOMATERIALS

Four basic molecular plans were developed through evolution over 3 billion years ago and are still used by all living things today. Modern cells use proteins, nucleic acids (such as DNA), polysaccharides, and lipids for nearly all tasks. A handful of other small molecules are specially synthesized for particular needs, but the everyday work of the cell is performed by these four basic plans. Of course, in bionanotechnology we are not forced to stay within these existing plans, but there are many advantages to exploring them first. Most notably, we can use the thousands of working natural bionanomachines as a starting point to build our own practical nanotechnology.

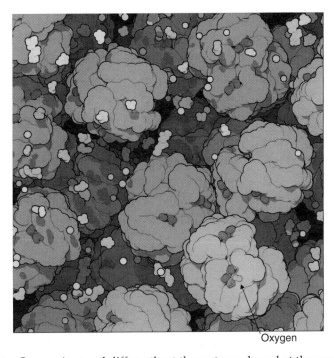
Oxygen

Figure 2-1 Oxygen is stored differently at the meter scale and at the nanoscale. At the meter scale, we store oxygen in high-pressure tanks. The oxygen is delivered into and out of these tanks in a continuous stream through tubes. The flow is controlled by smoothly machined valves. In contrast, at the nanoscale we transport oxygen molecule by molecule instead of in bulk. In red blood cells, the protein hemoglobin stores large amounts of oxygen at body temperature and without the need for high pressure. Individual oxygen molecules encounter hemoglobin by random diffusion, binding tightly when they meet. A complex shift in the orientation of the four subunits, mediated by the precise mating of atoms along the interface between subunits, allows hemoglobin to increase the gain on the interaction. This allows hemoglobin to gather oxygen efficiently when levels rise and to discharge all of the oxygen when levels drop.

Most Natural Bionanomachines Are Composed of Protein

Protein is the most versatile of the natural biomolecular plans. Protein is used to build nanomachines, nanostructures, and nanosensors with diverse properties. Proteins are modular, constructed of a linear chain of amino acids that folds into a defined structure, as shown in Figure 2-2. The longest protein chain (thus far) is titin with over 26,000 amino acids, and peptides with less than a dozen amino acids are used as hormones for cell signaling.

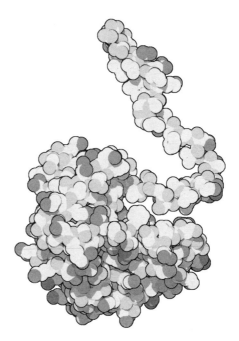

Figure 2-2   Proteins are constructed as chains of amino acids, which then fold into compact globular structures.

Typical soluble proteins have chains in the range of about 200 to 500 amino acids.

Amino acids are composed of a central α-carbon atom with three attachments: an amino group, a carboxylic acid group, and a side chain. Each successive amino acid is connected through an amide linkage between the amine of one amino acid and the carboxyl of the next amino acid in the chain. The amide linkage is rigid, strongly preferring a planar conformation of the four amide atoms and the flanking carbon atoms. The rigidity of the amide group is essential for the construction of nanomachinery with defined conformations. The rigid amide limits the number of conformations available to the chain. A more flexible chain, like the strings of aliphatic carbon atoms used in many plastics, is able to adopt many compact conformations of similar stability instead of forming a single folded structure with the desired conformation.

The combination of the rigid planar group and the exposed hydrogen and oxygen atoms gives rise to a limited range of stable conformations of the chain. Two conformations, shown in Figure 2-3, are particularly stable. They combine minimal strain and overlap in the molecular structure with a maximal number of hydrogen bonds between the exposed amide atoms. The first is the α-helix. The chain winds like a spring so that each amide oxygen interacts with the hydrogen atom three linkages down the chain. The second is the β-sheet, composed of several adjacent strands. Each strand is fully extended, and several strands bind side by side, forming a ladder of hydrogen bonds in between.

The chemical diversity of the different side chains provides the real ad-

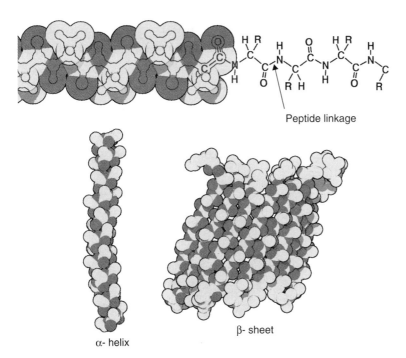

α- helix

β- sheet

Peptide linkage

Figure 2-3 The peptide linkage connecting amino acids contains a hydrogen bond donor, the H–N group, and a hydrogen bond acceptor, the O=C group. The remaining carbon in the protein chain carries a hydrogen and one of 20 different side chains, shown with an R here. Two conformations of protein chains, the α-helix and the β-sheet, are particularly stable, because the chain is in a relatively unstrained position and all of the possible hydrogen bonds between the amide groups are formed. This β-sheet, taken from the bacterial protein porin, has alternate strands running in opposite directions.

vantage of proteins as a structural material, allowing them to be used for many different functions. The 20 side chains (shown in Figure 2-4) used in natural proteins are chemically and structurally diverse. By arranging them in the proper order, the structure of the protein may be shaped and stabilized. Then particularly reactive side chains may be placed at key locations to perform the desired function.

A variety of modified amino acids are also used for specialized tasks. Some, like selenocysteine, are added directly to protein chains as they are synthesized, using alternate translations of the normal genetic code. Most, however, are created by modifying the natural 20 amino acids after they are incorporated into proteins. For instance, a hydroxyl group may be added to proline, which allows additional levels of hydrogen bonding that are important in the structure of collagen. In blood clotting proteins, an additional carboxylic acid group is added to glutamate amino acids, allowing them to bind more tightly to calcium ions.

The error rate of biological protein synthesis limits the size of individual chains that may be constructed consistently and accurately. In bacterial

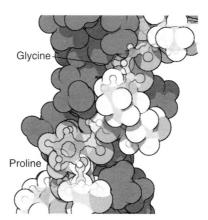

Figure 2-4A   Glycine and proline play special structural roles. Glycine is the smallest amino acid, with no side chain. Because it lacks a side chain, the backbone is not as constrained, making the protein chain more flexible at sites that incorporate glycine. It is used in regions that require tight conformational turns that are not possible for other amino acids and in crowded regions with strong steric blocking constraints, such as in the tight collagen triple helix shown here. Proline is the only cyclic amino acid, with two covalent bonds to the protein backbone. It forms a rigid kink in the protein chain. In collagen, this kink allows the chain to adopt a tight triple helix.

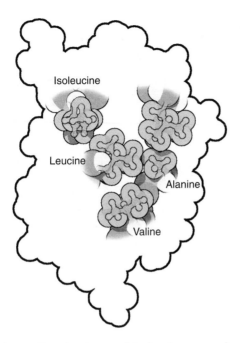

Figure 2-4B Alanine, valine, leucine, and isoleucine are carbon-rich amino acids with a range of sizes and shapes. They are relatively inflexible and strongly favor environments sheltered from water. Often, these hydrophobic residues drive folding of protein chains. The collection shown here are on the inside of insulin, forming a closely packed cluster inside the protein. Note that a variety of other short-chain carbon-rich chains are possible in this size range, such as a two-carbon chain and straight chains of three or four amino acids. However, only the four variations included here are genetically encoded in natural organisms.

cells, the genetic sequence is misread in about 1 in 2000 amino acids, substituting an improper amino acid at that location in the chain. However, these occasional errors are often tolerated and the misplaced amino acid has little effect on the function of the protein. However, processivity errors, in which synthesis of the protein terminates early and produces a truncated chain, are more serious. The frequency of processivity errors has been estimated at about 1 in 3000 amino acids. In response to these intrinsic limits, average protein chains fall in the range of 200–500 amino acids, although spectacular exceptions, such as the muscle protein titin, are synthesized for specialized tasks.

We can find examples of proteins everywhere we look. Most proteins are soluble structures, performing their jobs in solution. Egg white exempli-

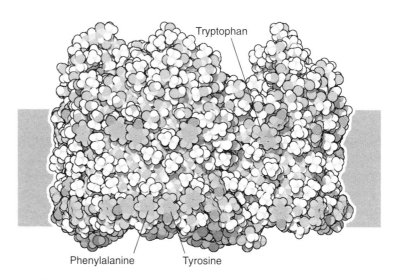

Figure 2-4C Phenylalanine, tyrosine, and tryptophan have large aromatic side chains. They favor environments sheltered from water, and, along with the carbon-rich amino acids shown in Figure 2-4B, they drive the folding of protein chains. These rings often stack on top of one another or on top of DNA bases and are used to provide specificity for aromatic rings binding in active sites. Tyrosine is a special case, with an aromatic phenyl ring and a hydroxyl group at the end. This provides a perfect mix of properties for interacting with small organic molecules, so tyrosine is often used in protein binding sites both to stabilize the carbon-rich portions of a ligand and to hydrogen bond with the ligand. Porin, a bacterial protein that spans a lipid membrane, is shown here. The membrane is shown schematically as the dark stripe. Note how these aromatic amino acids are arranged around the perimeter of the molecule, forming a belt that interacts with the carbon-rich membrane.

fies the macroscopic properties of a concentrated solution of soluble proteins: a viscous solution that denatures, turning opaque, when heated. Freeze-drying yields a deliquescent powder, which for many proteins may be dissolved in water to yield an active protein. Large protein biomaterials are also built. The rubbery material in tendons is largely composed of the protein collagen, and the tough but flexible material of hair and fingernails is largely composed of the protein keratin. These proteins are extensively cross-linked for additional strength.

Bionanotechnology is exploiting the potential of proteins in every way imaginable. Powerful methods for creating custom proteins are available, as described in Chapter 3. The major current limitation is basic knowledge. We

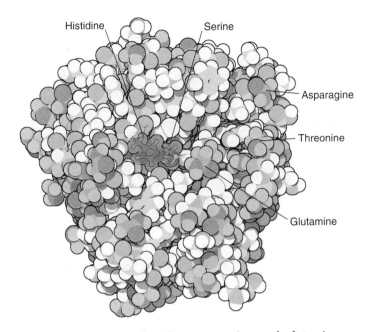

Figure 2-4D Serine, threonine, histidine, asparagine, and glutamine are amino acids with diverse hydrogen-bonding groups. They are very common on protein surfaces, where they interact favorably with the surrounding water. They are often used to glue protein structures together and to form specific interactions with other molecules. Histidine is a special case. It contains an imidazole group, which may adopt neutral and charged forms under slightly different conditions. In the neutral form, it combines a protonated secondary amine, which is electrophilic and may donate a hydrogen bond, with a tertiary amine, which is strongly nucleophilic and can accept a hydrogen bond. Histidine is used infrequently in proteins, being incorporated mainly for specialized catalytic tasks. For instance, it is being used here in the protein-cutting enzyme trypsin to activate a serine amino acid. Normally the hydroxyl group on serine is unreactive, but when activated in the proper environment it is an effective catalysts for reactions that require addition or abstraction of hydrogen atoms. Histidine also coordinates strongly with metal ions and is used to construct specific metal-binding sites.

need to understand and be able to predict the processes by which proteins fold into their stable, globular structure.

Nucleic Acids Carry Information

Nucleic acids are modular, linear chains of nucleotides, ranging up to hundreds of millions of nucleotides in length. Two forms are commonly used:

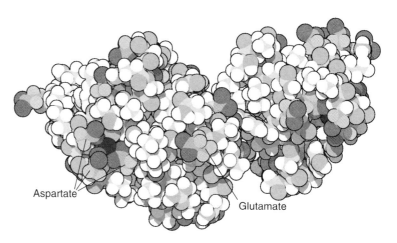

Aspartate

Glutamate

Figure 2-4E Aspartate and glutamate contain a carboxylic acid group. Under biological conditions of neutral pH, these residues are ionized with a negative charge. They are common on protein surfaces, are widely used in chemical catalysis, and bind tightly to metal ions. Calmodulin uses three acidic amino acids to hold a calcium ion, and many others are scattered on the surface where they interact with the surrounding water.

ribonucleic acid (RNA) and deoxyribonucleic acid (DNA). DNA differs by the absence of a single hydroxyl group in each nucleotide, making it slightly more stable under biological conditions. Nucleic acid chains are far more flexible than protein chains, so nucleic acids adopt a wide range of conformations. The structure is largely determined by the interactions of the bases in each nucleotide. Because they are aromatic, they stack strongly on top of one another in water solutions. Also, the bases have been chosen for their ability to interact specifically with one another through a coded set of hydrogen bonds. The combination of strong stacking interaction and specific lateral hydrogen bonding leads to the familiar double helix structure for DNA and RNA (Figure 2-5).

Four bases are commonly used to construct DNA: adenine, guanine, cytosine, and thymine. In RNA, the similar uracil base replaces thymine. The four bases have very similar chemical properties and differ primarily in the arrangement of hydrogen-bonding acceptors and donors around their edges. Two canonical pairings—adenine with thymine and guanine with cytosine—are strongly favored in typical double helices. Many other pair-

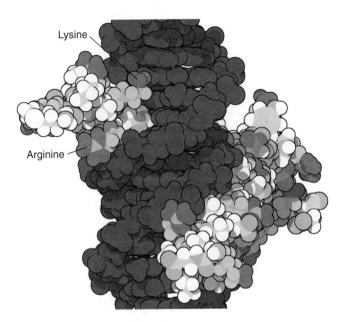

Lysine

Arginine

Figure 2-4F   Lysine and arginine contain basic groups at the end of long, carbon-rich chains. The amine at the end of lysine and the guanidinium group of arginine are both ionized under biological conditions and carry a net positive charge. They are found primarily on the surface of proteins and are widely used for recognition of negatively charged molecules. In particular, arginine is important in the binding of proteins to nucleic acids, as seen in this repressor protein bound to a DNA double helix. The long, flexible carbon-rich portions of these side chains also play a role in interaction with other carbon-rich molecules.

ings are also possible, and in special cases modified bases are used to expand the repertoire of base pairing interactions.

The uniform chemical properties of the nucleotides limit the functions of nucleic acids. They are specialized for applications in nanoscale information storage and retrieval. Each nucleotide encodes two bits of information. Information is duplicated and read through specific interactions of each nucleotide with a specific mate. Despite these limitations, the ribosome, which is perhaps the most important molecule in the cell, is composed predominantly of RNA.

We rarely encounter pure nucleic acids in daily life. When isolated from cells and dried, nucleic acids are fibrous, appearing much like cotton fibers. But bionanotechnology is extending the utility of nucleic acids far past stor-

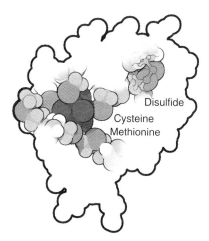

Figure 2-4G   Cysteine and methionine contain sulfur atoms. Cysteine is the most reactive of the amino acids, with a thiol group. Cysteine is important in its ability to form covalent disulfide cross-links, linking two cysteine residues in different portions of the protein chain. Cysteine is also used much like serine in chemical catalysis. Cysteine coordinates strongly with metal ions and is used to form specific metal-binding sites. Methionine has a hydrophobic sulfur atom. It is often used like the carbon-rich amino acids, to promote the folding of proteins. The sulfur atom is also nucleophilic and coordinates with several types of metal ions. The small electron-carrying protein ferredoxin shows many of these uses of cysteine and methionine. A disulfide linkage is seen at upper right, and four cysteines hold a cluster of iron and sulfur (shown in gray) at the center. Two methionines embrace the cluster, further stabilizing it inside the protein.

age of genetic information, as described in Chapter 6. Because of the strong, predictable pairing of bases, large structures may be created by designing the appropriate sequence of bases and then allowing double helices to form. Nucleic acids, despite their limited chemical diversity, are also starting to be harnessed for jobs normally performed by proteins, such as chemical catalysis and biosensing.

Lipids Are Used For Infrastructure

Surprisingly, some of the largest structures built by cells are composed not of large macromolecules like proteins or nucleic acids but instead by a fluid aggregate of small lipid molecules. The lipids used by living cells have been designed to aggregate into a defined set of useful infrastructures. They are

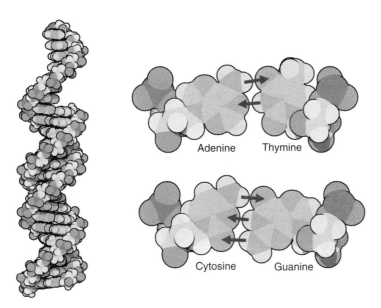

Adenine   Thymine

Cytosine   Guanine

Figure 2-5 A DNA double helix is shown on the left. Each strand is composed of a backbone composed of sugars and phosphates and bases that are stacked inside. Genetic information is stored and transmitted through a coded set of hydrogen bonds between bases, as shown on the right. Adenine pairs with thymine, forming two hydrogen bonds, and guanine pairs with cytosine, forming three hydrogen bonds. The result is a four-letter code capable of storing two bits of information per nucleotide.

small molecules that combine two different chemical characteristics into a single molecule. They are composed of a polar or charged group, which interacts favorably with water, attached to one or more carbon-rich chains, which strongly resist dissolving in water. This dual character causes them to act much like protein chains when placed in water. As described more fully in Chapter 4, lipids self-organize into globules or membranes, with all of the charged/polar groups facing water and all of the carbon-rich tails packed inside (Figure 2-6).

A few natural lipids are used for different applications in cells. Of course, these are just the starting points for bionanotechnology: Many variations are possible on the theme. The most common natural lipids are phospholipids and glycolipids. These are constructed around a central glycerol molecule, which has three hydroxyl groups allowing attachment of three separate groups. Two are typically attached to fatty acids: A carboxylic acid attaches to the glycerol, and the long carbon chain, typically between 16 and

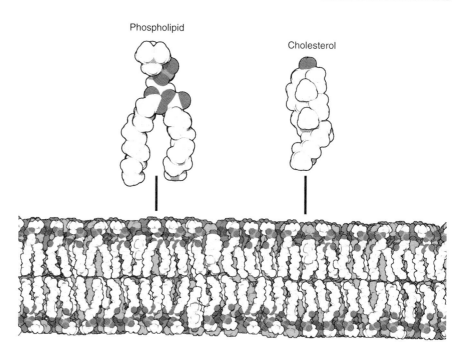

Figure 2-6 Lipids are used to build membranes that resist the passage of large molecules and ions. Here, a bilayer is seen in cross section and is composed of phospholipids and cholesterol, enlarged at the top. Note the dual chemical nature of the lipids, with carbon-rich portions shown in white and charged/polar portions in red.

24 carbon atoms long, extends away. Several unsaturated bonds may be incorporated into the fatty acid to form rigid kinks that are used to modify the character of the aggregates formed. The remaining position of the glycerol is taken by the water-soluble group, which may be a phosphate group or other charged/polar group.

Cholesterol and other sterols are built with a different plan. They use a rigid, bulky lipid molecule, composed of many fused hydrocarbon rings, that is about as long as the carbon chains attached to phospholipids and glycolipids. A hydroxyl at one end is hydrophilic, aligning cholesterol in the membrane. Cholesterol is added to membranes in varying amounts to modify their characteristics. Because cholesterol is rigid, it tends to inhibit the motion of neighboring lipids, reducing the fluidity of the membrane and also making it less permeable to small molecules.

Lipids are widely used for cellular infrastructure, forming the membranes that surround cells and the organelle compartments inside. They are impermeable to ions and to larger polar molecules, from sugars to proteins. Carbon-rich molecules, however, pass freely through these membranes. This is why alcohol disperses rapidly through the body, crossing all barriers.

Polysaccharides Are Used in Specialized Structural Roles

Polysaccharides are the most heterogeneous of the four molecular plans. Sugars, the building blocks of polysaccharides, are covered with hydroxyl groups. The polymers are created by connecting the hydroxyl groups together, offering many possible geometries for polymerization. In nature, many different linear and branched polymers are constructed for different needs (Figure 2-7). For instance, the simple sugar glucose is found in several forms. When attached with a ($\beta1\rightarrow4$) linkage, glucose forms a long, straight chain that is used for structural fibers in cellulose, such as in the tough fibers of cotton. However, if a slightly different ($\alpha1\rightarrow4$) linkage is used, the chains form tight coils, forming powdery starch granules. Branched chains are also commonly used for specific functions, attaching new chains at multiple points on a single sugar branch point. Glycogen is an example: It is a dendrimer composed of increasingly branched glucose chains. It is used for storage of glucose, so the tight dendrimeric form is compact and presents many free ends for removal of individual sugars when needed.

The many hydroxyl groups in polysaccharides form hydrogen bonds with other hydrogen bond donors and acceptors, offering two modes of interaction. In some cases, individual polysaccharide chains associate with a large volume of water, forming thick solutions or a gluey gel. In this form, carbohydrates coat most of our cells, forming a gluey, protective coat. The glycoproteins in mucus will give you an idea of their properties. In other cases, carbohydrate chains associate tightly side by side, aligning hydroxyl groups and forming strong fibers with little water trapped inside. In this form, polysaccharides are used for large-scale infrastructure and energy storage. Some of the most impressive biological creations, including sturdy tree trunks and tough, waterproof carapaces in arthropods, owe their strength to polysaccharides.

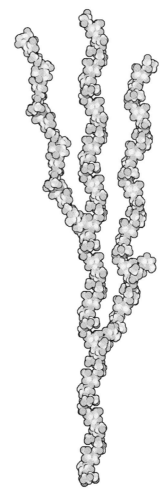

Figure 2-7   Polysaccharide chains often have a branched structure and are covered with water-soluble hydroxyl groups. Because of their strong interaction with water, they tend to form extended, disordered structures.

THE LEGACY OF EVOLUTION

If we were given the task of designing a living cell, we probably would not take the parsimonious approach seen in nature. Think, for a minute, about the machines that we design in our everyday world. A computer contains microscopically patterned silicon chips, an injection-molded plastic body, metal wires that carry electric current, and phosphorous compounds coated onto glass that are bombarded with electrons to produce light. Each of these components is constructed with a different process, according to a different set of plans, often in a different part of the world. Cells are more uniform—they use only a handful of synthetic techniques and rely on a few simple molecular plans to build their many different bionanomachines. This can be both an asset and a liability. Biological molecules have their limitations—they require water environments with the proper temperature, pH, and salinity. So why has nature limited biomolecules to these particular plans?

The process of evolution by natural selection is the root cause. Evolution places strong constraints on the form that biological molecules adopt, strongly favoring *modification* over *innovation*. Evolution proceeds through the passing of genetic information from generation to generation. At each step, small changes may be introduced, so that children are different in some small way from their parents. But it is essential to make *small* changes. If a change compromises a single one of the multifold processes of life, the children will die. Cells and organisms must maintain a living line all the way back to the earliest primordial cells. If a single generation fails to create a living descendent, all of its biological discoveries will be lost.

Evolution is far more limiting than the technology of our familiar world. If we create a computer that doesn't function, perhaps while testing a new type of computer chip or keyboard button, we can scrap it and go back to the drawing board. We have lost some time and money, nothing more. But if a critical molecular component is changed in a cell, it must be right every time, or the cell pays the price of extinction.

Of course, evolution proceeds despite these dire consequences, as evidenced in the diversity of modern life forms. Cells have several levels of redundancy within which to experiment with new molecular machines. First, the blueprints for a given protein may be duplicated within the genome. Then the duplicate may be modified without regard to its original function, as long as the original is still there. Gene duplication is very common in the evolution of life—our own DNA is filled with examples. For instance, about 200 million years ago, the gene encoding hemoglobin, the protein that carries oxygen in the blood, was duplicated. This allowed a second form of hemoglobin to be optimized for a different function, while the original continued with its job in the blood. The new hemoglobin gradually acquired a stronger affinity for oxygen, binding it more tightly than the normal blood hemoglobin. Today, this specialized hemoglobin is used in the blood of a fetus, so that it can capture oxygen from the mother's blood.

Sheer numbers also aid evolution. Cells rarely live all alone. Typically, a colony of bacteria or a herd of cattle is the biologically relevant entity. Within this population, there is ample room for experimentation. Occasional lethal mutations may be tolerated, as long as the rest of the population survives. Individuals with rare improvements may then dominate in later generations. Slowly, these differences cause the populations to diverge, forming new varieties of organisms and ultimately creating entirely new species.

Human immunodeficiency virus (HIV) demonstrates the power of populations and the progress possible through evolutionary optimization. When HIV reproduces, it uses an enzyme to make copies of its small genome. This enzyme is error-prone, making far more mistakes than the similar enzymes that copy our own DNA. This may seem like a problem, but it actually gives HIV a great advantage. In an infected individual, 10 billion viruses are made every day. Many of these will have mutations somewhere in their genome, many of which are ultimately lethal. But the population is so large that there are always a few normal viruses to carry on, and occasionally one of the mutants is better than the original virus. For instance, when a person is treated with anti-HIV drugs, the normal virus is killed but some of the mutants are able to survive. Within weeks, powerful drug-resistant strains dominate the population. This is evolution in action, but accelerated to rates far greater than the slow pace normally seen in nature. Natural populations of higher organisms take hundreds or thousands of years to make evolutionary changes, because of the high accuracy of their DNA-copying mechanisms and long lifetimes. HIV, on the other hand, may shift in a matter of days, by using its sloppy copying enzyme and its large population of individuals.

The hallmark of biological evolution is the plasticity provided by mutation and gene duplication. A great number of variants are tested within a population, slowly improving and optimizing every component. The many amazing machines described in this book are testament to this plasticity. Evolutionary optimization allows the design of subtle mechanisms that are difficult with our familiar "rational design" approach to engineering. For instance, proteins often incorporate a complex range of flexibility into their function, using small shifts in local structure to grip targets and complex motions to modify and control activity. Furthermore, these motions are not transitions between a few rigid states, like the on and off states of a switch. Instead, these motions are optimized in the context of the constant, random thermal motion induced by the watery environment, so that these motions must be thought of as structural ensembles of many functional conformations. This is a great challenge for rational nanoscale design, requiring full knowledge and description of the entire conformational range accessible to each state. For evolutionary design, however, this is an easy task. Evolution makes lots of changes and keeps whatever works. Evolution does not design anything before starting construction—instead, it builds many, many prototypes.

Evolution, however, carries with it an important constraint: the problem of legacy. Once a key piece of machinery is perfected and placed in use, it is difficult to replace it or make major modifications without killing the cell. This is particularly true of central molecular processes like the reading and use of genetic information, the production of energy, and reproduction, all of which require the concerted effort of dozens of complex molecular machines. This leads to a remarkable uniformity in all earthly living things when observed at the molecular level. All are built of the same basic components, discovered once by evolution and used in all subsequent organisms.

Of course, in our own bionanotechnology, we are not restricted by evolution. We are free to create and test any nanomachine that we can imagine. We are not constrained by the *mechanism* of evolution. However, we are currently limited to the *materials* of biomolecular evolution. To use the principles of structure and function perfected by biological evolution, we must start with biomolecules and change them carefully into our own bionanomachinery. And, as described in Chapter 6, evolution itself, after being suitably accelerated in the laboratory, is now being harnessed to design bionanomachinery.

Evolution Has Placed Significant Limitations on the Properties of Natural Biomolecules

Biomolecules have evolved to act ideally under biological conditions, which are relatively mild compared with the conditions often endured by our macroscale machinery. This imposes significant limitations on the range of possible functions and environments for bionanomachinery. Nearly all biomolecules are designed for function in water solution. Most are designed for function at 37°C, neutral pH, and weak but significant salinity. In special cases modifications may be incorporated to extend the range of stability, but the basic limitations of covalent, organic molecules and self-assembly, described in Chapter 4, place hard limits on the range of conditions tolerated.

One last caveat has been noted by Eric Drexler. For the most part, biological evolution has produced nanomachines with relatively short life spans. Most proteins last only days. This is a different approach than that typically taken (as most consumers would hope) in macroengineering. We typically build things to last, whereas bionanomachinery is typically built to perform a single task. Even structures that one might feel are permanent, such as bones, are continually disassembled, repaired, and rebuilt. This is a different paradigm than that used in macroengineering. It is wasteful of energy, requiring constant regeneration of resources, but is perfectly tailored for the constant sensing and response to environmental conditions that are a hallmark of life. Perhaps this paradigm of planned obsolescence, in the context of complete recycling, will provide a useful model for creation of human artifacts in an environmentally responsible manner.

GUIDED TOUR OF NATURAL BIONANOMACHINERY

Nature has already realized many of the dreams of nanotechnology. Thousands of bionanomachines have been selected and perfected by evolutionary optimization to perform nanoscale tasks accurately, consistently, and under specific control. These bionanomachines use all of the engineering tricks used in our familiar macroscale machines: construction from many tight-fitting parts, hinges for bending, rotating axles and bearings, digital information storage, chemical adhesion and chemical power. The examples on the next few pages (Figures 2-8 through 2-16) will give you an idea of the diversity of these amazing, molecule-sized machines.

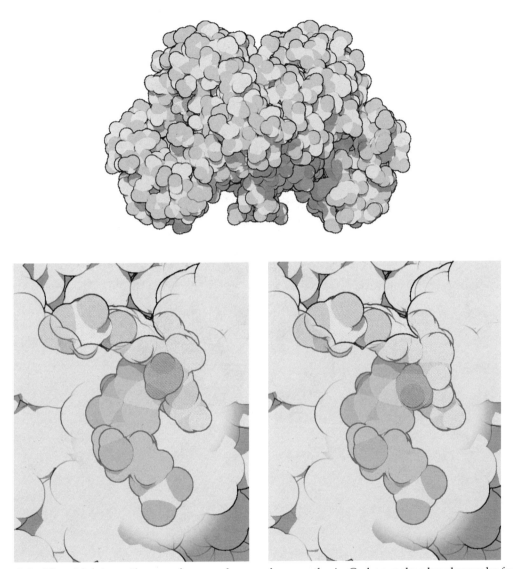

Figure 2-8 Thymidylate synthase performs carbon mechanosynthesis. Carbon-carbon bonds may be forged with special molecular tools. It is not sufficient merely to press the carbon atom against its target. To perform carbon mechanosynthesis efficiently under mild conditions, the carbon atom must be activated and the target site must be prepared for acceptance. The enzyme thymidylate synthase performs a specific carbon mechanosynthesis reaction, placing a new methyl group on a nucleotide base. The methyl group, shown here in red, is activated by bonding it to a cofactor molecule, shown in pink. The cofactor is carefully designed to carry carbon but to be more stable without it. When carbon is transferred, the carrier pops into a more stable form, preventing the carbon atom from being transferred back. Thymidylate synthase precisely aligns the target molecule and the activated carbon atom, as shown in the illustration, and forcibly performs the transfer. Proximity is not sufficient; instead, the enzyme must tailor the perfect environment for transfer. The target molecule is surrounded by chemical groups that shift its electronic structure to favor the transition.

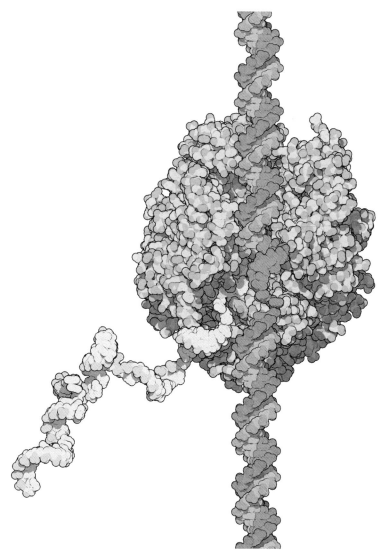

Figure 2-9 DNA carries a library of information. Biological information is stored at remarkable density. A single bacterial cell, barely a micrometer in largest dimension, stores 70 Kbyte of information in its genome. A typical compact disk uses a similar space to store a single bit of information. Biological information is stored in a form that is chemically stable and redundant for ease in repair. This medium is used in one mode to store blueprints for construction and in a second mode to control synthesis. Every aspect of the structure of DNA is used to carry information. Inside the double helix, the genetic information is stored by using a specific set of hydrogen bonds. In this illustration, RNA polymerase (shown in gray) is copying the information from the DNA strand (shown in red) into a temporary messenger RNA strand (shown in pink). As described in Chapter 4, the surface of the DNA helix is also used to carry information on the regulation and storage of DNA. This "extragenetic" information is read by proteins that wrap around the helix. Even the stiffness of the DNA helix is used to control the location of molecules that interact with the DNA.

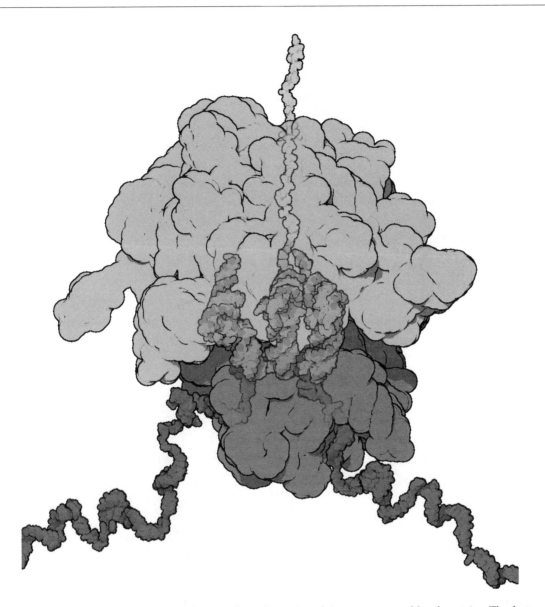

Figure 2-10 Ribosomes are complete factories for information-driven nanoassembly of proteins. The factory performs a modular assembly, reading information in a linear storage medium and arranging 20 different modules into a linear chain. Any length and any sequence of modules may be created at will, simply by creating the appropriate set of instructions. Ribosomes are fully general: Any protein may be created using a standard set of starting and stopping instructions and a standard coding scheme for the blueprint of the desired product. The ribosome, shown here in gray, is composed of two parts that trap the RNA message strand that is read, which is shown in red. The small subunit on the bottom positions the RNA message, and the large subunit on the top performs the synthesis reaction, expelling the new protein through a hole.

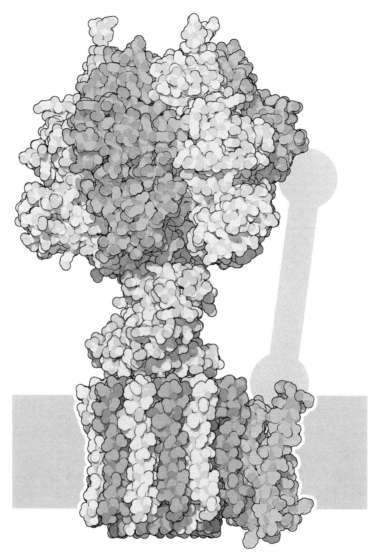

Figure 2-11 ATP synthase is a rotary motor and generator. ATP synthase performs an energy transformation, converting electrochemical energy into chemical energy and vice versa. At the bottom is a motor driven by electrochemical gradients, which is bound inside a lipid membrane (the membrane is shown schematically as a gray stripe). This motor is composed of a rotor composed of a cyclic ring of proteins (shown in gray) and a stator (shown in pink). The stator guides the flow of hydrogen ions across the membrane and transforms it into motion of the rotor. At the top is a chemically powered motor, driven by the breakage of the unstable molecule ATP. This motor is composed of a ring of six proteins (in pink) with an eccentric axle through the center (in gray). Cleavage of ATP forces a change in the shape of the surrounding proteins, driving rotation of the axle. The whole complex may be used in either direction. The electrochemical motor can drive the chemical motor, creating ATP in the process, or the chemical motor can be powered by breakage of ATP, turning the electrochemical motor and creating a gradient.

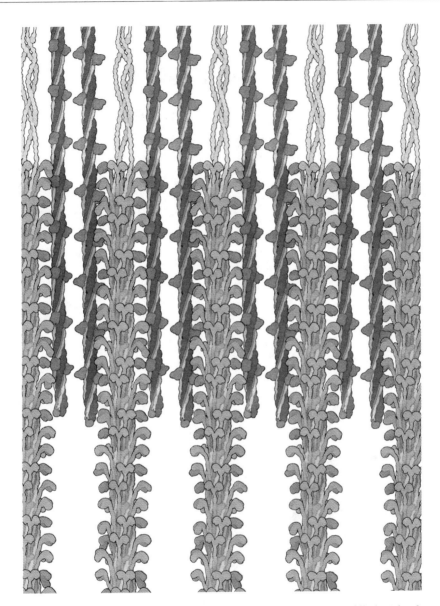

Figure 2-12 Actin and myosin form an engine of contraction. Muscle cells are filled with a huge array of interdigitated myosin filaments and actin filaments. A small section is shown here, with myosin filaments in pink and actin filaments in gray. Chemical energy is converted into mechanical work by myosin. The many myosin heads climb along the neighboring actin filaments, powered by ATP. In a contracting muscle, each myosin head may perform a power stroke five times a second, moving along the actin filament about 10 nm with each motion. About 2 trillion myosin power strokes are needed to generate the force to hold a baseball in your hand, but your biceps have a million times this many, so only a fraction of the myosin in a muscle is exerting force at any given time.

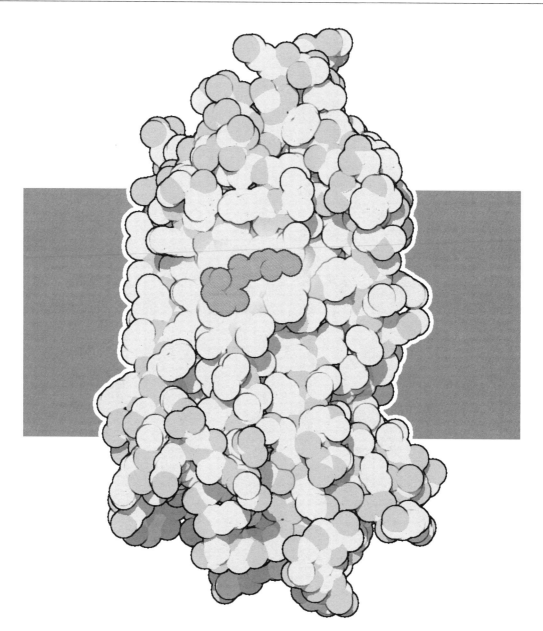

Figure 2-13 Opsin is a light sensor. Because biomolecules interact by intimate association of their surfaces, they easily sense subtle changes in surface conformation. Therefore, most biosensors transduce a signal, such as light or chemical conditions, into a shift of surface shape. Opsin contains the light-sensitive molecule retinol (shown here in red) buried inside the protein chain. As described in more detail in Chapter 5, retinol absorbs a photon of light, flipping one kinked *cis* bond into a straighter *trans* conformation. This change in shape is amplified by the surrounding protein. The resultant shift in protein shape is easily sensed by proteins inside the cell, which begin a cascade of responses, ultimately resulting in a nerve signal to the brain.

Figure 2-14 As discovered by Buckminster Fuller, triangular modules may be used to construct large, sturdy structures. Perfectly symmetrical triangles form only icosahedra, but larger structures may be built by allowing a small amount of flexibility at each point of attachment. Clathrin uses this principle at the nanoscale level to create a reversible packaging and delivery system. Three-armed triskelion molecules (one shown in red) form a transient cage on membrane surfaces (shown in pink), pulling out a vesicle filled with molecular cargo. Note that the triskelions have formed hexagonal and pentagonal arrangements in the geodesic network. The flexibility of the triskelions allows formation of a variety of spherical and ovoid shapes.

Figure 2-15 Nanomedicine was discovered a billion years ago and continues to protect our bodies from disease and infection today. The immune system contains hundreds of biomolecules that selectively seek out invaders and destroy them. In this figure, many Y-shaped antibodies (shown in pink) are attacking an HIV particle. Note that the surrounding blood serum includes hundreds of other antibodies, each designed to bind to a different foreign molecule.

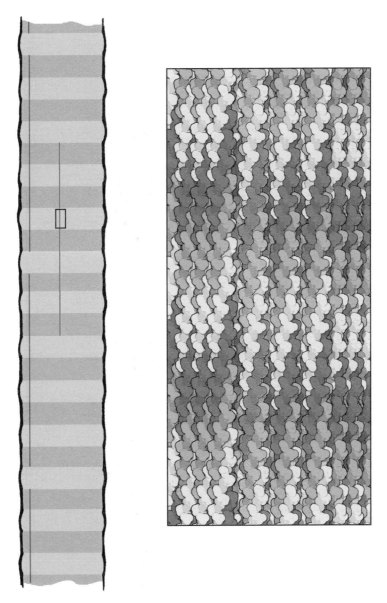

Figure 2-16 Sturdy, resilient natural biomaterials are built by all higher organisms. This molecule is collagen, the most plentiful protein in your body, which forms long cables that strengthen tendons. The individual molecules are long, thin rods composed of three tightly wound protein strands. The neighboring strands then pack side by side, forming a strong fiber. Many hydrogen bonds and cross-linking bonds between strands add to this strength. By incorporating small crystals of apatite between fibers, the material can be made even tougher, forming bones that may outlast the life of the organism by many years.

BIOMOLECULAR DESIGN AND BIOTECHNOLOGY

3

*The problems of chemistry and biology can be greatly
helped if our ability to see what we are doing, and to do
things on an atomic level, is ultimately developed—a
development which I think cannot be avoided.*
> —Richard Feynman

Today, we have an abundant variety of methods for doing things on an
atomic level. Chemists were already constructing molecules atom-by-atom
at the time that Richard Feynman gave his visionary talk, and today, chemistry is a powerful method for constructing molecules with several dozen
atoms. In the time since Feynman's talk, the fields of physics and biology
have yielded additional methods for working at the atomic scale. Physicists
are pushing atoms around with atomic force microscopes and trapping
them with optical tweezers, and biologists have harnessed the rich collection of natural bionanomachinery to perform our own custom molecular
tasks.

Bionanotechnology is widely accessible, more so than any other cut-
ting-edge application of nanotechnology. Silicon-based fabrication techniques, to reach the nanometer scale, must push the resolution of fabrication machinery to their limits, making the process expensive and available
only to large corporations and laboratories with extensive resources. The diamondoid models of molecular nanotechnology are purely theoretical. But

Bionanotechnology: Lessons from Nature. By David S. Goodsell
ISBN 0-471-41719-X   Copyright © 2004 John Wiley & Sons, Inc.

powerful tools for designing bionanomachines are available to anyone with a computer and imagination, and effective tools for producing these custom bionanomachines are accessible to any moderately-sized biotech start-up company.

Current methods of biotechnology excel at modification. This is a powerful capability that leverages the extensive body of working nanomachinery that is available from natural sources. We can introduce specific changes into the plans for a given protein, or we can splice together the plans for several different proteins, creating a hybrid molecule with combined function. Using these modified plans, we can then engineer bacteria to produce large quantities of the mutant or chimeric protein. Thousands of academic and industrial laboratories are using these methods for medicine, bioremediation, and countless other applications. And several exciting new techniques based on biological evolution, described in Chapter 6, allow thousands of modifications to be tested simultaneously, greatly speeding the discovery of biomolecules with new functions.

Design of entirely new bionanomachines, on the other hand, is currently more difficult than modification of natural bionanomachines. Evolution has designed complex machines with subtle mechanisms, incorporating flexibility and self-assembly in ways that are difficult to predict and design. Designing bionanomachines from scratch is currently a great challenge that is under intensive study in many laboratories. Ideally, we want total control. For instance, we might want to build a "nanotube synthase" that constructs carbon nanotubes of defined size and geometry. We would like to be able to go to our computer and design a protein that would fold into a stable structure, creating an active site that performs this chemical reaction. Unfortunately, there are gaps in our knowledge that must be filled before this capability is possible. Today, we cannot reliably predict the folded structure of a protein from its chemical sequence, and, given a folded structure, we cannot consistently predict its chemical activity. But these two steps are currently under scrutiny by scientists, with the firm expectation that they will be solved in the foreseeable future. Then, true biomolecular design will be a reality.

This chapter presents an overview of the many techniques that are available for the design, synthesis, and analysis of biomolecules. This information is by no means comprehensive and provides only an introduction to these powerful methods. Many excellent workbooks and recipes are available for each of these methods.

RECOMBINANT DNA TECHNOLOGY

Recombinant DNA technology is the core capability of bionanotechnology. This technology allows us to construct any protein that we wish, simply by changing the genetic plans that are used to build it. Two natural enzymes—restriction enzymes and DNA ligase—are the keys to recombinant DNA technology, allowing us to edit the information in a DNA strand (Figure 3-1). Before the discovery of these enzymes, researchers modified the genetic code of living organisms by using biology's own tools of mating and

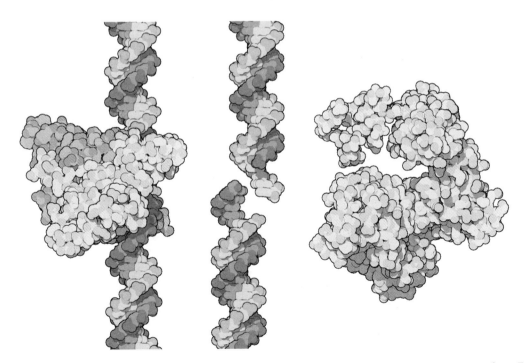

Figure 3-1 Recombinant DNA technology relies on two key enzymes. Restriction enzymes, such as *EcoRI* shown on the left, cut DNA at specific sequences. Often, these enzymes make a staggered cut, producing "sticky ends," as shown in the center. DNA ligase, shown on the right, connects two strands back together.

crossing or by random mutagenesis with chemicals or ionizing radiation. Today, researchers modify the genetic code rationally at the atomic level.

Restriction enzymes are unbelievably useful enzymes (I am reminded of the babble fish in Douglas Adams's *The Hitchhiker's Guide to the Universe*). They are built by bacteria to protect themselves from viral infection. The bacterium builds a restriction enzyme that cleaves DNA at one specific sequence. At the same time, it protects its own DNA by modifying the bases at this same sequence, so the restriction enzyme does not cleave its own genome. Invading viral DNA, however, is instantly chopped up by the restriction enzyme, because it is not similarly protected. Serendipitously, many restriction enzymes make staggered cuts in the two DNA strands, instead of cutting both strands straight across the DNA helix. Here is where biotechnology steps in with a new use for these enzymes. These ends are "sticky" and readily associate with other sticky ends of similar sequence. So restriction enzymes may be used to cut DNA, producing sticky ends that may be pasted back together in custom orientations. Thus restriction enzymes, originally evolved merely for their destructive capacity, are now tools for atomic-precision editing of large pieces of DNA.

Today, recombinant DNA technology has flowered. Clever researchers are continually discovering new methods for harnessing the protein production machinery of the cell in new ways. Consistent methods, often in the form of commercial kits, are available for every possible process. We can find and extract specific genes from organisms. We can duplicate and determine the sequence of large quantities of these genes. We can mutate, recombine, and splice these genes or create entirely new genes nucleotide by nucleotide. Finally, we can replace these genes into cells, modifying their genetic information.

DNA May Be Engineered with Commercially Available Enzymes

Customized DNA is routinely created in thousands of laboratories worldwide. Together, biological and synthetic techniques allow the construction of large DNA strands composed of natural DNA sequences or entirely new DNA sequences. A successful service industry has arisen that provides basic expertise for DNA manipulation. You can readily purchase stretches of DNA of any given sequence and all the enzymes needed to handle them.

Researchers use a wide variety of natural biomolecules for handling DNA. Well-characterized protocols and commercial sources for these enzymes are available, so these processes are available to any modest laboratory. A few of the most important are:

(1) *Restriction enzymes* are isolated from bacteria. Over 100 types are available commercially. Each one cuts DNA at a specific sequence of bases. Typically, restriction enzymes are composed of two identical subunits, so they attack DNA symmetrically and cut at specific palindromic sequences.

(2) *DNA ligase* reconnects broken DNA strands. When two sticky ends anneal, DNA ligase is used to reconnect the breaks.

(3) *DNA polymerase* creates a new DNA strand by using another strand as a template, creating a double helix from a single strand. It is used to fill single-stranded gaps and to copy entire pieces of DNA.

Chemical synthesis of DNA perfectly complements these natural biomolecular tools for manipulating DNA. Current methods allow the automated synthesis of DNA strands about 100 nucleotides in length. Two complementary strands are easily constructed and annealed in solution to form a double helix. Short oligonucleotides are routinely synthesized and are available commercially.

Once a new DNA is constructed, large quantities are produced by two major methods: DNA cloning and the polymerase chain reaction. The term "cloning" refers to the creation of identical copies without the normal processes of sexual reproduction: copies of mice or sheep, identical cultures of cells, or, in this case, many identical copies of a particular fragment of DNA. In *DNA cloning*, a bacterial cell is used to create many identical copies of a DNA sequence. One method is to insert the DNA sequence of interest into a virus, which then infects bacterial cells and forces them to make many copies. Alternatively, a bacterial plasmid may be used. Bacteria naturally contain small circles of DNA—plasmids—in addition to their main genome. To clone a DNA sequence, we add it to a bacterial plasmid and then insert it into bacteria. The plasmid is then copied each time the bacterium divides, forming large quantities of the DNA as the culture grows (Figure 3-2).

The *polymerase chain reaction* (PCR) is a method for copying a small sample of DNA. It takes advantage of an efficient, heat-stable DNA polymerase isolated from bacteria that live in hot springs. As shown in Figure 3-3, PCR proceeds in cycles, doubling the number of DNA strands at each step. PCR is so powerful you can start with a single strand of DNA and get as much as needed out.

Once engineered DNA strands are built, we need methods to use them to create custom proteins. Proteins are conveniently made in engineered cells using *expression vectors*, plasmids that contain the gene specifying the protein along with a highly active promoter sequence. The promoter, which is often taken from a virus, directs the engineered cell to create large quantities of messenger RNA based on the plasmid DNA in the vector. The cell then synthesizes the protein based on this messenger RNA. Bacteria are the most widely utilized host cells that are engineered for protein production. Engineered bacteria create large amounts of protein, often comprising 1–10% of the total cellular protein. Also, bacteria are easy to grow, and inexpensive fermentation methods allow growth of high densities of bacterial cells with modest resources.

However, bacteria present several significant limitations. Animal and plant cells often modify their proteins after they are synthesized, and bacteria do not perform these modifications. In particular, many animal and plant proteins have carbohydrate groups attached to their surfaces, and bacteria do not add these groups to engineered proteins. This can be a fatal problem in the production of proteins for use in medical treatment. Many of these proteins must have the appropriate carbohydrate groups to be active, and the immune system can react dangerously to improper carbohydrate groups (for instance, the need to be careful of blood types during transfusions is due to differences in the carbohydrates attached to cellular proteins). Engineered yeast cells, insect cells, or cultured mammalian cells may be used in cases where the proteins must be modified for proper action.

Another problem with engineered bacteria, which is occasionally an

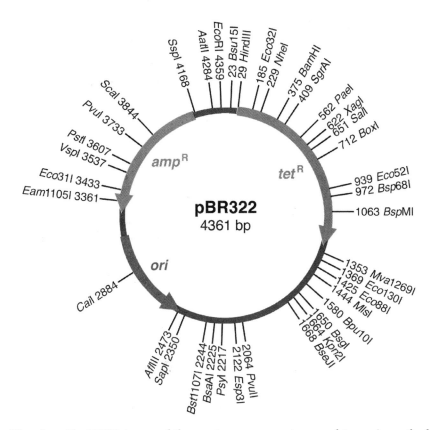

Figure 3-2 The plasmid pBR322 is one of the most common vectors used to engineer the bacterium *Escherichia coli*. A map of the plasmid, which contains 4361 base pairs of DNA, is shown here. The plasmid contains a region that directs the replication of the plasmid (*ori*) and two genes that encode proteins for antibiotic resistance, one for ampicillin (*amp*^R) and one for tetracycline (*tet*^R). The sites that are cleaved by different restriction enzymes are shown surrounding the circle. By choosing the appropriate enzyme, the plasmid can be cut at specific locations. Researchers add new genes to the plasmid by cutting at one of the restriction sites and splicing in the new DNA. The drug-resistance genes provide a clever method of determining whether or not any bacteria have taken up the plasmid. For instance, if the new DNA is added at the *Pst*I site at position 3607, the inserted DNA will disrupt the ampicillin-resistance gene. Thus bacteria that contain this new plasmid are easily identified and separated from bacteria that do not contain the plasmid: They will be resistant to tetracycline but sensitive to ampicillin.

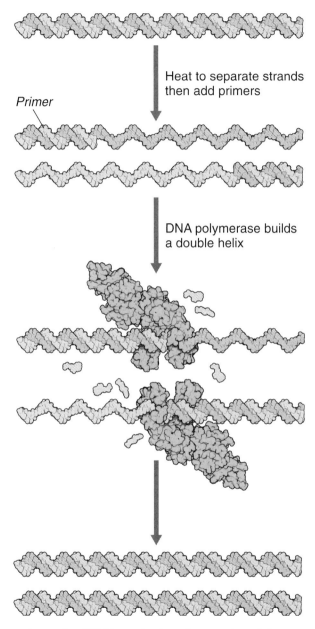

Figure 3-3   Through repeated rounds of DNA synthesis and separation of the two strands, the polymerase chain reaction amplifies the amount of DNA in a sample. (1) The process begins with a single strand of DNA. (2) It is separated by heating, and short primer strands are added to the ends. (3) DNA polymerase builds a new strand using the separated strands as a template. (4) At the end of the cycle, there are two identical DNA double helices. This cycle is repeated, doubling the DNA at each step. The use of a heat-stable polymerase is the trick to making this an automated process, because it can survive the heating step of each cycle.

asset, is that the proteins tend to aggregate when they reach high concentrations, forming inclusion bodies. Inclusion bodies are dense aggregates of proteins that are easily visible in the microscope, often extending entirely across the bacterial cell. They are formed when new proteins associate randomly before they can undergo the proper folding process. Inclusion bodies are extremely tough, and harsh conditions must be used to solubilize the individual protein chains. In many cases, the purified proteins may then be folded under conditions that lead to the proper structure. If it is possible to renature the functional protein from inclusion bodies, they can be a substantial aid to purification. Because inclusion bodies are denser than most of the other structures in the cell, they are easily separated from the other cellular components simply by centrifuging the cell extract.

Proteins may also be created without the help of living cells, by isolating the protein production machinery and performing the reactions in the test tube. The first step of protein production, the transcription of DNA into a messenger RNA, is now routine with purified RNA polymerase. However, the second step, the synthesis of proteins based on purified messenger RNA in cell-free systems, is still a technical challenge. In some cases, extracts of the cell cytoplasm, containing the protein synthesis machinery along with everything else, are effective. Extracts can, however, encounter problems with limited energy supply and the presence of protease and nuclease enzymes that cleave the products and RNA message. Specialized continuous-flow cell-free systems have been developed to overcome this problem.

Attempts to recreate protein synthesis with purified preparations of the components have also been successful. But because of the complexity of the system, requiring over 100 separate components, they are still limited to relatively modest yields. These methods are primarily used in research rather than industrial production of proteins. The advantages, however, of cell-free protein production make it an attractive goal. It provides a controlled method for synthesizing proteins that are difficult in engineered bacteria, such as membrane-binding proteins, proteins that are toxic to bacteria, and proteins that include unusual amino acids. Development of efficient cell-free translation mechanisms is an area of active research.

Site-Directed Mutagenesis Makes Specific Changes in the Genome

In many cases, we might want to make a few small changes to an existing natural protein, to tailor its function for a given application. Site-directed mutagenesis is used in these cases to modify the amino acid sequence of a protein by making specific changes in the existing gene encoding it. In this way, we can make atomically precise changes in the structure of a protein, altering structure and function. A wide variety of methods are available for modifying existing genes. Some of these methods are so reliable that prepackaged kits are available from commercial sources.

Site-specific mutations are conveniently introduced into existing genes with specially designed oligonucleotides, as shown in Figure 3-4. These short strands match the normal sequence of the DNA except at the point where the change is desired. The change may be a single amino acid change or a short insertion or deletion. Once the change is made, cloning and expression is used to construct the modified protein.

Site-directed mutagenesis has revolutionized molecular biology. It is extremely powerful for determining the function of specific amino acids or regions within a protein. For instance, individual amino acids may be mutated one at a time, looking for those that compromise the function. In this way, the active site of an enzyme or the binding site of a hormone may be localized. Site-directed mutagenesis is also widely used in attempts to improve the stability of proteins, by engineering in cross-linking residues or improving the fitting of residues within the protein interior (Figure 3-5). These methods are humbling, however. All too often, we discover how difficult it is to predict modifications that do not disrupt the stable structure and function of natural proteins.

Fusion Proteins Combine Two Functions

Recombinant DNA techniques are also used to combine entire genes, forming a larger fusion protein that combines the functionality of all of the pieces. Special care must be taken when designing the linkage site, so that the fused proteins will not block one another when folding into their active structures. Fortunately, many natural proteins are very robust and perform their functions even when fused to another large structure.

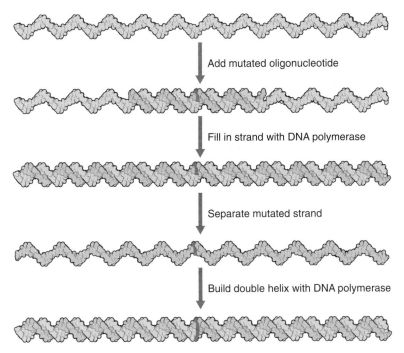

Figure 3-4 In site-directed mutagenesis, specific changes are incorporated into genes by using specially designed small oligonucleotides. The oligonucleotide matches the gene, except for the place where the change is desired. To make the change, the short oligonucleotide is annealed to the strand under conditions that allow pairing despite the mismatch at the desired site. DNA polymerase is then used to fill in the rest of the DNA sequence, using the short oligonucleotide as the primer. This engineered strand is then separated, and the original DNA strand is discarded. The result is a strand complementary to the original DNA, but with changes in the region where the oligonucleotide was bound.

Fusion proteins can harness the natural delivery mechanisms in cells. In our cells, proteins are targeted to different compartments (such as the mitochondria and endoplasmic reticulum) through the use of a short *signal peptide* the end of the protein chain. These peptides are used as handles to recognize the proper location for the protein and are later clipped off after the protein is delivered. Recombinant DNA techniques can be used to attach signal peptides to any given protein, specifying its location. For instance, the signal peptide for secretion may be attached to a protein of interest. This modified protein will then be released into the surrounding medium, ready for harvesting and purification.

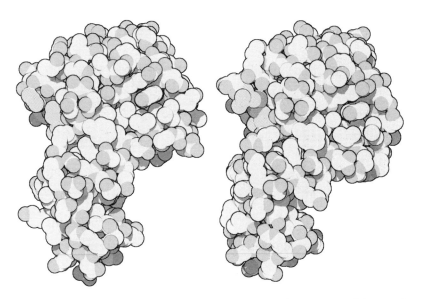

Figure 3-5 The enzyme lysozyme has been extensively engineered in search of ways to improve its function and stability. The native enzyme is shown here on the left, with two amino acids at opposite ends of the protein chain shown in pink. When the protein folds, these two amino acids end up close to one another in the structure. In one engineered version of lysozyme, shown on the right, these two amino acids have been changed to cysteine. When the engineered protein folds, the two cysteines form a disulfide bond, shown in red, that stabilizes the folded structure.

Chimeric proteins have also shown great utility. Two proteins with different functions are combined, creating a hybrid protein with both functions. For instance, anticancer immunotoxins have been created by combining an antibody that binds to cancer cells with a toxin that kills the cell (Figure 3-6). The immunotoxins seek out cancer cells and kill them, reducing side effects of normal cancer chemotherapy. For research applications, the green fluorescent protein from jellyfish has been attached to many proteins to study the location of these proteins within living organisms. Portions of the organism that are making the protein will glow green.

MONOCLONAL ANTIBODIES

Many applications in bionanotechnology, such as biosensing and recognition of disease in nanomedicine, require an effective method for recognizing individual molecules. Fortunately, the immune system of animals is de-

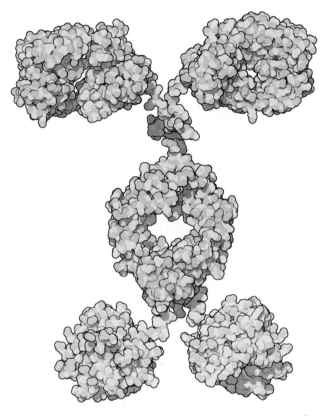

Figure 3-6 Immunotoxins are created by fusing a toxic protein, shown here in pink, to an antibody, shown in gray. They are being tested for use in cancer therapy.

signed to perform exactly this function, so we can look to the immune system for methods. The central tools of the immune system are antibodies (Figure 3-7). Antibodies are proteins that specifically bind to molecules that are foreign to the organism, such as infecting pathogens. Our immune system is capable of creating 10^{15} different types of antibodies, each with a different binding specificity. By combining this natural library of molecules with modern methods of antibody production, it is now routinely possible to obtain antibodies capable of high-affinity recognition of virtually any molecule.

Antibodies are built by B-cells (a type of white blood cell) and comprise about 20% of the protein in human blood serum. Each B-cell builds only a single type of antibody. On their surfaces, B-cells display a tethered version of their particular antibody. These tethered antibodies link the specificity of

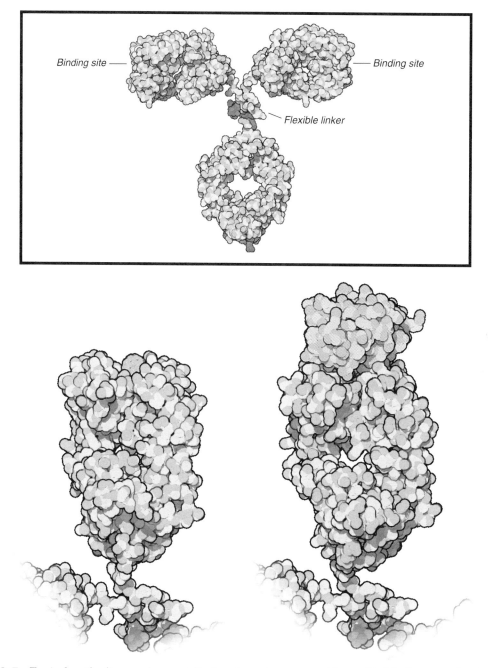

Figure 3-7 Typical antibodies, as shown at the top, have two arms with specific binding sites for their targets connected by a flexible linker to the central domain. The shape of antibody binding site is designed to match the target molecule. Two examples are shown at the bottom. The antibody on the right has a large, flat surface that binds to the protein lysozyme (shown in pink), and the antibody on the left has a deep pocket that binds to buckminsterfullerenes.

the antibody to the genetic instructions needed to build it. When the antibodies on the surface of a given B-cell bind to a target, the B-cell proliferates and creates large quantities of the soluble antibody with the same specificity. As the cell proliferates, it can also modify the antibodies made by daughter cells. Daughter cells with improved binding characteristics will then be selected for further growth, and those with reduced binding ability will be removed.

Because of their strong binding to specific molecules of interest, purified antibodies are used in many applications. They provide a ready handle for recognizing a given target. For instance, specific antibodies are used to recognize HIV in blood, providing the means for testing for HIV infection, and hormones in urine, providing the means for pregnancy tests. Antibodies can be used to localize specific proteins within organisms, with applications in research and medicine. A few of these applications, including immunotoxins and catalytic antibodies, are discussed in more detail in Chapter 6.

If animals are immunized with a given molecule (termed an antigen), they will produce a variety of antibodies that bind at different locations on the target. In most cases, this heterogeneous mixture of antibodies is not useful, and we desire a single, uniform antibody with the desired characteristics. We need to isolate the single B-cell that produces the desired antibody and then to propagate it in culture. Unfortunately, B-cells have limited life spans when grown in culture, so that the desired quantities of antibody cannot be consistently produced. The solution to this problem was developed in the late 1970s. The antibody-producing cells are fused with an immortal cell line—a line of cells taken from a tumor that will grow continuously in culture. The fused cell will grow in culture and will produce the antibody. This is termed a "monoclonal" antibody, because it is produced by a clone of identical fused cells. In practice, monoclonal antibodies may be raised against nearly any target with current techniques.

BIOMOLECULAR STRUCTURE DETERMINATION

As noted by Richard Feynman, the key to understanding in biology is the ability to see what cells are doing. Our understanding of the mechanics of biomolecular function, and our ability to engineer them for new functions, entered a new era when the first atomic structures of proteins were determined. In the late 1950s, John Kendrew and colleagues solved the structure of myoglobin, revealing in breathtaking atomic detail how protein chains can be used to store oxygen. Since then, experimental techniques have been perfected and thousands of protein, nucleic acid, lipid, and polysaccharide structures are available. This priceless resource is available to the public at the Protein Data Bank (http://www.pdb.org).

Biomolecular structure determination, although significantly streamlined since its beginning, is still an expensive endeavor in terms of resources and expertise. In the sections below, I give a short overview of the major methods used for determination of biomolecular structures. I focus on discussion of the utility and limitations of the final structures obtained by these methods, instead of the methods themselves. When using these structures as a starting point for bionanotechnology, it is essential to understand how accurately these structures represent the actual structure of the molecule.

X-Ray Crystallography Provides Atomic Structures

X-ray crystallography currently provides the most detailed information on atomic structure (Figure 3-8). The equipment and expertise needed to obtain crystals, collect X-ray data, and solve the structure is substantial and typically requires a dedicated laboratory with considerable resources. Many excellent references are available describing the theory and methods of this fascinating discipline.

The experimental information obtained from a crystallographic analysis is a three-dimensional map of electron densities. This map shows the observed density of electrons at each point in the crystal lattice. The resolution of this map—the spacing of points at which the electron density is resolved—is dependent on the quality of the crystals. The best crystals of biomolecules provide very high-resolution data, and features separated by less than an angstrom (0.1 nm) will be easily resolved in the electron density maps. Typical crystallographic studies of proteins are poorer, in the range of 1.5- to 3.0-Å resolution. At 1.5-Å resolution, individual atoms are easily distinguished, but at 3.0-Å resolution, knowledge of the covalent structure is needed to interpret the less well-resolved contours.

The researcher then interprets the electron density in terms of an atomic

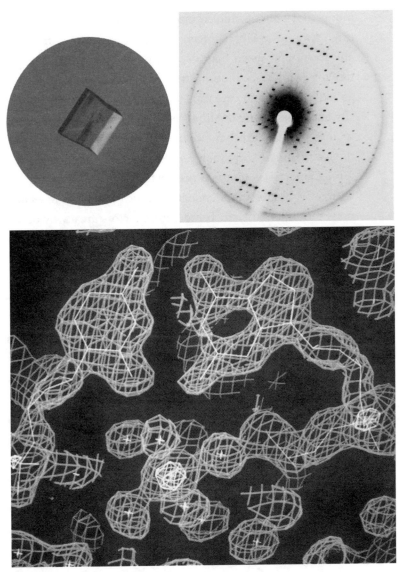

Figure 3-8 X-ray crystallography begins by growing a crystal of a pure molecule, as shown at upper left. Then the crystal, which may only be a fraction of a millimeter in size, is placed in an intense beam of X rays. The crystal diffracts the X rays into a characteristic pattern of spots, at upper right. This pattern is then analyzed in the computer to yield an electron density map, shown at the bottom, that reveals location of all of the electrons in the crystal. The map shown here is part of a DNA crystal. Regions inside the contours are dense with electrons. You can see a cytosine-guanine base pair near the top and below it a calcium ion surrounded by water molecules. This map of electrons is then interpreted to determine the location of each atom in the structure.

model. For data of high resolution, this yields a high-quality atomic structure. Atomic positions may be determined to within a fraction of an angstrom, and users may be confident when using the structures. When moving to lower-resolution structures, at 3.0 Å or worse, the structure is not as well defined and care must be taken when interpreting the electron density and when using the final coordinates. Mobile areas of the structure and regions on the surface may not be well resolved in the electron density, so the resultant structure may represent only a single interpretation of the observed data. The temperature factors (B-values) of the atomic positions within the model are often a good indication of the level at which the positions should be trusted. Temperature factors are a way of modeling the disorder of each atomic position. Model atoms are often treated as a Gaussian distribution around an atomic center, with the B-value controlling the width of the bell-shaped curve. B-values of about 10 represent atoms with sharply defined positions, whereas values of 30 or higher represent highly disordered atoms that must be treated with caution.

The need for crystals is a major limitation of X-ray crystallography. Because the biomolecule is bound within a perfectly ordered lattice, the structure represents only a snapshot of the conformations that may be relevant

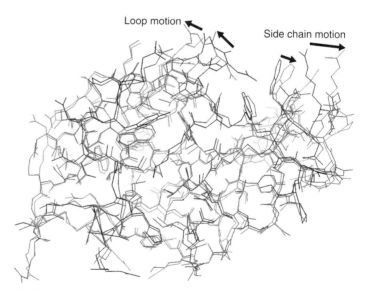

Figure 3-9 Structures of a given protein may show differences when analyzed in different crystals. Two different structures of the protein lysozyme are superimposed here, one shown with red bonds and one with black bonds. Note the differences in the exterior loops of the chain and the locations of the side chains.

when the biomolecule is free in its natural environment. In many soluble proteins, this limitation is not prohibitive. Several proteins have been studied in several different crystal lattices, revealing very similar overall structures. Functional aspects of biomolecular flexibility, however, must be studied with multiple crystals obtained under varied conditions (Figure 3-9).

NMR Spectroscopy May Be Used to Derive Atomic Structures

Nuclear magnetic resonance (NMR) spectroscopy is the workhorse for determining molecular structure in chemistry. Data from NMR spectroscopy characterizes the local environment of atomic nuclei inside molecules. Certain atomic nuclei have an intrinsic magnetic moment that aligns in a strong magnetic field. This alignment may be perturbed by a radio frequency pulse of appropriate wavelength, and when the nuclei relax to their aligned state, they emit characteristic radio frequency radiation that reflects the local environment of the atom. The characteristic NMR spectra have been used ex-

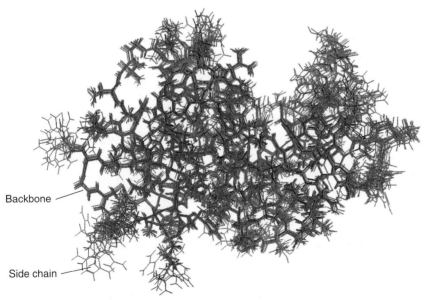

Backbone

Side chain

Figure 3-10 The data from NMR spectroscopy are often interpreted to yield an ensemble of possible structures. This picture shows an ensemble of 10 different lysozyme structures that are all derived from a single data set. Note how the main protein chain is very similar in each structure, because of the large amount of data that specifies its structure. The side chains are less well defined and adopt a range of conformations. Compare this ensemble with the changes observed in the two crystallographic structures of lysozyme in Figure 3-9.

tensively in chemistry to define the covalent and conformational structure of small organic molecules. They are now being used to study larger molecules, such as small proteins and nucleic acids.

For biomolecules, the spectra get very complex, so more elaborate NMR techniques are needed to allow study of the many similar atoms in the molecule. Two-dimensional NMR techniques currently allow determination of structures of small proteins. In this technique, multiple radio frequency pulses are used to perturb multiple nuclei. If one nucleus is excited, it will modify the absorption and emission of nuclei in the immediate vicinity. Ultimately, these small shifts are used to develop a list of nuclei that are in close proximity in the molecule. Refinements in these methods have extended the range of NMR to small and medium-sized proteins, with 100–250 amino acids.

NMR experiments identify the distances between nuclei that are spatially close to one another and the local conformation of atoms bonded together. To determine the structure of the entire biomolecule, these local pieces of information must be combined into an atomic model. Most often, the data provide a list of constraints, tabulating pairs of atoms that are close to one another and conformations of given bonds in the structure. The researcher then develops an atomic model that is consistent with the list of constraints. Often, the results of an NMR analysis are presented as an ensemble of structures, such as that in Figure 3-10, each of which fit the constraints. This ensemble may be interpreted in two ways. It might represent the range of conformations that the molecule might adopt when free in solution, or it might represent a range of structures, one of which is the actual structure. A mixture of these two interpretations is probably closest to the reality.

Electron Microscopy Reveals Molecular Morphology

Electron microscopy has a long and venerable history in all aspects of nanoscale science. It is perhaps the most intuitive approach to imaging macromolecular objects, because it is so similar to light microscopy. Theoretically, electron microscopes should be able to see subatomic structure, but practical limitations—imperfections in the magnetic optics and prob-

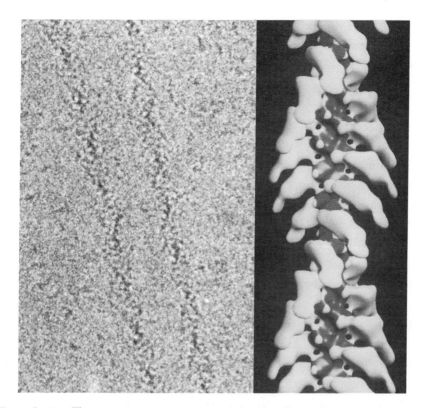

Figure 3-11   Electron microscopy can reveal the overall morphology of a biomolecule. A complex of actin with the motor domain of myosin, studied by Ron Milligan at the Scripps Research Institute, is shown here on the left. Note the low contrast between the two strands and the surrounding ice—try squinting to see the overall shape better. A computer-generated reconstruction based on the frozen image is shown on the right. By combining many portions of the electron micrograph, the reconstruction creates an averaged three-dimensional model of the molecule.

lems with specimen preparation, contrast, and radiation damage—limit the resolution to about 2 nm for biomolecules. This provides enough detail to determine the overall morphology of biomolecules and biomolecular complexes, but not to see individual atoms.

Electron microscopy provides information that is not available from any other experimental sources and is often used to study assemblies that are too large for other methods or molecules that undergo structural changes under different conditions (Figure 3-11). When information from electron microscopy is combined with atomic information from X-ray crystallography, NMR spectroscopy, and molecular modeling, the atomic structure of large assemblies may be constructed. Our understanding of the ribosome, many large viruses, and the interaction of actin and myosin in muscle cells has greatly benefited from this combined approach.

Both transmission electron microscopy and scanning electron microscopy are used to determine the structure of bionanomachinery. Transmission electron microscopes are similar to light microscopes: The electron beam illuminates a thin sample, and the microscope determines the relative transparency of different regions. The contrast of biological specimens is often very low, so they are often stained with salts of heavy metals, such as uranium or osmium. Unfortunately, this staining procedure can introduce artifacts during treatment and drying. Cryoelectron microscopy reduces these artifacts but introduces problems with contrast. The sample is frozen in ice, so the contrast between the biomolecule and the surrounding ice is low. Often, structures are determined by analysis and averaging of many individual particles to build up an averaged image of the molecule that reduces the noise introduced by the low contrast. In the best cases, electron tomography can provide a three-dimensional image of the molecule. Images are collected from the sample tilted at a range of angles, and differences between the tilted samples are used to construct the three-dimensional model.

Scanning electron microscopy provides a three-dimensional image by looking at electrons that are scattered or emitted from the surface of the specimen. The sample is prepared by fixing and drying and then is coated with a thin layer of metal. The specimen is then scanned with a narrow beam of electrons to image the surface. The resultant images are very intuitive, giving a good feeling for the three-dimensional contours of the specimen. But, because of the need for a metal coat, the resolution is often much lower than in transmission microscopy, at about 10 nm. This is sufficient for images of large assemblies, such as the arrangement of actin and myosin in entire muscle sarcomeres.

Atomic Force Microscopy Probes the Surface of Biomolecules

Atomic force microscopy, developed in the early 1980s, is a newcomer relative to the other techniques described here. The approach is more akin to touch than to vision. A sharp probe is scanned over the surface of the sample, recording the height at each point and yielding a topographic map of the surface. Of all of the methods for determining molecular structure, this provides the most direct connection between our world and the atomic world.

To scan the surface, the sample is moved in a rasterlike pattern under the tip and the sample is raised and lowered to apply a constant force on the tip. Both motions—the lateral scanning and the changes in sample height—are controlled by piezoscanners, and forces on the cantilever are detected by shining a laser beam on the back of the cantilever and watching for motions in the reflected beam. The sample may be scanned in a constant contact mode, in which the tip is always in contact with the sample. This allows very accurate measurements of height, but the high shear forces as the tip is forcibly scanned across the sample can be problematic for soft biological samples, which are often attached only weakly to the sample surface. The tapping mode of scanning solves these problems. The tip is oscillated such that the tip just touches the surface during the scan. Because the contact is very short, shear forces are reduced. The resolution of the image is dependent on the sharpness of the tip and is typically in the range of 5–10 nm.

Atomic force microscopy became a powerful tool for study of biological molecules when methods were developed to analyze samples in water instead of using dried samples. Dried biological samples retain a thin layer of water on their surface. When the probe is scanned over the surface, capillary forces can dominate the interactions between the tip and the sample, masking the dispersion/repulsion forces that define the shape. The tapping mode improves this somewhat but must be operated at an amplitude that is sufficient to break the capillary interaction with each oscillation. Today, however, these problems are solved by simply immersing the entire sample and tip in solvent. Capillary forces are removed, and the tip-sample interaction reflects only the shape of the molecule.

Atomic force microscopy has shown great success in imaging bionanomachinery. Numerous systems, from individual DNA strands to entire chromosomes, have been imaged at near-atomic level. The advantage of the technique is the use of conditions similar to those encountered in cells, so that the molecules are in conformations that are appropriate to their natural function. The technique has also shown great success in nonimaging applications, in which the microscope is used to measure forces between molecules or forces as biomolecules are stretched and unfolded. The microscope

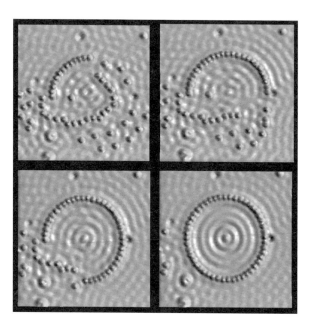

Figure 3-12 At IBM, atomic force microscopy has been used to arrange individual iron atoms into a circular "corral," allowing study of the unusual quantum mechanical properties of the arrangement. (Figure from http://www.almaden.ibm.com/vis/stm/corral.html)

provides a sensitive method for measuring forces along the trajectory of stretching or separation and has yielded insights into protein folding, DNA conformation dynamics, and enzyme specificity.

Scanning force microscopy has been used, in a number of spectacular demonstrations, to make specific atomic level changes in molecules, such as the arrangement of argon atoms to create quantum corrals with peculiar quantum mechanical characteristics (Figure 3-12). Similar applications with biomolecules have been scarce. Fred Brooks and coworkers have attempted to push viruses into position on surfaces. Perhaps the most powerful approach for the future will be the attachment of specific functionalities to atomic force microscopy tips.

MOLECULAR MODELING

In designing new bionanomachines, computation works hand in hand with experimentation, often in an iterative manner. Molecules are designed in the computer, and the best ideas are synthesized and tested. The lessons learned are then applied in the next round of computational design, and so on. This approach has been highly successful in rational drug design, most notably leading to many of the powerful drugs used to treat AIDS. Computation also often allows exploration of systems that are experimentally inaccessible, providing predictions and directing further research and development.

Bionanomachines Are Visualized with Computer Graphics

Computer graphics revolutionized the study of biomolecules and now is an indispensable tool for all of the molecular sciences. Computer graphics allows us to visualize the unfamiliar shapes, properties, and interactions of molecules in a manner that is familiar and intuitive. Molecular graphics provides the first window onto a new project, allowing the researcher to explore and understand the molecules that will be built or modified.

Because molecules are orders of magnitude smaller than the wavelength of visible light, direct imaging of individual molecules is not possible. Therefore, various representations have been developed, as shown in Figure 3-13. The best representations capture the key properties of the molecule in a visual form, presenting us with a three-dimensional model that we can comprehend, but in such a way that the properties of the visual model relate directly to the nanoscale properties of molecule.

Today, computer graphics hardware and software are sufficiently fast to allow interactive representation of even the largest biomolecules. Excellent commercial and free software is available for visualizing molecular structures. Some of the most popular packages include:

(1) *RasMol*. A compact, self-contained program for the display of molecular structures. A flexible scripting language allows choice of representation styles and coloration and selective display of portions of a molecule.
(2) *Protein Explorer*. Another compact molecular display program.
(3) *Chime*. A Java plug-in for display of molecular structure within HTML pages.

Links to many of these programs are available on-line at the Protein Data Bank (http://www.rcsb.org/pdb/software-list.html).

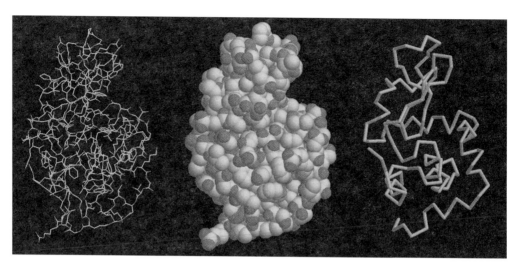

Figure 3-13 Lysozyme is shown in three common representations, as drawn in the popular viewing program RasMol. On the left is a bond diagram, showing every covalent bond linking atoms in the structure. In the center is a spacefilling diagram, representing each atom as a sphere. On the right is a ribbon diagram, a schematic representation of the topology of the protein chain. Note the advantages and disadvantages of each representation. Bond diagrams allow exploration of the detailed geometry of the molecule but are often too complicated for easy comprehension. Spacefilling diagrams give a good feeling for the shape and size of the protein. Ribbon diagrams are excellent for understanding biomolecular folding and topology.

Computer Modeling Is Used to Predict Biomolecular Structure and Function

Molecular modeling techniques allow the researcher to build any desired molecule based on the known molecular geometry of the component atoms. Molecular mechanics then applies a mathematical force field to this three-dimensional structure to define the interactions between each atom. The system may be used in several ways:

(1) *Optimization.* A crude molecular structure may be optimized, looking for a structure that best corresponds to the force field constraints.

(2) *Normal mode analysis.* The collection of forces and positions may be analyzed for specific harmonic modes, representing the major bending and twisting modes of the entire molecule.

(3) *Molecular dynamics.* A simulation of the molecule at a given temperature may be performed, following the molecule through time as thermal fluctuations are applied.

(4) *Free energy perturbation.* A transition is modeled by shifting the system smoothly from a given starting state to a different final state, following the process in detail. Often, a nonphysical path is used that is thermodynamically identical with the real transition but is more amenable to computational modeling.

Typical methods allow modeling of biomolecules of several thousand atoms, and typical molecular dynamics simulations may be run for nanoseconds. This is sufficient to look at events such as catalysis and local structural changes, but it is not sufficient for longer time scale processes such as protein folding and molecular docking, for which other methods are currently used (see below).

Many commercial and academic software packages are available for molecular modeling and molecular mechanics. Popular software includes:

(1) *Insight (BioSym).* Commercial package with excellent molecular modeling tools and diverse molecular mechanics methods.

(2) *Sybyl (Tripos).* Another commercial package with excellent molecular modeling tools and molecular mechanics methods.

(3) *Amber (UCSF).* Academic package with the full range of minimization, normal mode and dynamics simulations.

The Protein Folding Problem

A major hurdle must be crossed before bionanotechnology will have general applicability: We must be able to predict the folded structure of a protein starting only with its chemical sequence. Without this ability, we will merely shadow evolution, poking and prodding existing proteins until they are changed into something that we want.

The protein folding problem poses grave difficulties for two reasons. The first is the sheer magnitude of the problem. Typical proteins have several hundred amino acids. Each is connected to its neighbors through two flexible linkages that may adopt a range of stable conformations. In addition, each amino acid has a flexible side chain that can adopt a number of stable local conformations. Together, these many levels of torsional freedom define a staggeringly large conformational space that is beyond all current computational prediction methods.

The second problem lies in the method used to estimate the stability of each trial conformation during a prediction experiment. Folded proteins have thousands of internal contacts, each of which adds a tiny increment of stabilization to the entire structure. Many water molecules are freed as proteins fold, as the protein chains shelter their carbon-rich portions inside. This freeing of water is a strong force pushing proteins toward a folded structure. Entropy, on the other hand, works against the favorable energies of internal contacts and water release. Because of a decrease in entropy, a tightly organized protein globule is far less likely than a floppy, extended chain. The energy gains from contacts and released water are spent in forcing the chain into its compact form. Looking at the whole system, these two opposing forces just about cancel out, with a small excess on the favorable side. It is this favorable surplus of stabilizing energy that we must predict when trying to solve the protein folding problem, choosing the one folded structure with the most stable total energy. However, the value of this energy is calculated as the difference between two large sums, each of which may have significant errors.

Together, the large search space and the cumulative errors in scoring functions have thwarted many protein folding predictions. The most successful approaches have used simplified models, often approximating the protein chain on a lattice to reduce the space of conformations to search. These simulations, however, are still some distance from predicting accurate three-dimensional structures for use in functional design and prediction.

Currently, the best predictions of protein structure are obtained by homology modeling. For this, proteins are modeled based on the known structure of a similar protein. In an analysis of protein structures in the Protein Data Bank, proteins that are identical in about 30% of their amino acids (evaluated by aligning the sequences of the two proteins) were shown to have homologous structures. In these structures, the folding and topology of the protein chain are similar, but the local details in loop regions may differ. Homology modeling takes advantage of this observation. The structure of a new protein may be modeled based on the structure of a known protein with similar sequence (if one is available). Computer modeling is used to build structures for loops and to create coordinates for specific amino acids that are changed. For proteins with sequence homology of 60% and higher these models can be very accurate, and in the range of 30% to 60% the models can be useful for predicting the overall properties of the protein structure, such as identifying surface residues or looking for a global shape.

Algorithms for predicting the local structure of protein chains based on the sequence are also currently quite robust. These methods are calibrated by using the known structures of many proteins. They then they scan along a new protein sequence, classifying each region as α-helical, β-sheet, or other (for more on protein secondary structure, see Chapter 4) or classifying the regions as surface exposed or buried within the protein. These techniques make correct predictions for approximately 70% of the amino acids, enough to identify the basic folding pattern of the protein. A similar technique for identifying segments of protein chains that cross through membranes is particularly successful. The unusual chemical properties of these regions, because they interact with membranes instead of water, make them easily identifiable, so prediction algorithms are about 95% accurate.

Note, however, that both homology modeling and secondary structure prediction may have limited applicability in bionanotechnology. Both methods rely on the fact that the protein sequences under study are derived from evolutionarily optimized proteins. The researcher begins with the knowledge that the sequence adopts a stable, functional, folded structure. Many examples of single amino acid changes that entirely disrupt a structure are known, and these prediction methods typically would not be able to identify that type of localized problem. However, the success of homology modeling is an excellent place to begin for bionanotechnology. It provides a technique for the design of modified proteins, starting from existing stable protein folds and modifying step by step to add new functionality.

Docking Simulations Predict the Modes of Biomolecular Interaction

Specific interactions between molecules are the basis of most biomolecular processes. Enzymes recognize the shape of the molecules they modify and create an environment that promotes a chemical reaction. Antibodies have a binding site that perfectly matches its target. Proteins interact and communicate through specific binding sites. Methods for predicting these interactions are necessary for analysis and design in bionanotechnology.

Accurate, consistent methods are available for the prediction of the binding of a small molecule—a ligand or inhibitor—to a biomolecular target (Figure 3-14). The most successful methods combine two capabilities. First, they use a fast algorithm to search the many ways that the molecules can fit together. Second, they use an energetic model that accurately predicts the energy of interaction. The algorithm searches many possible binding

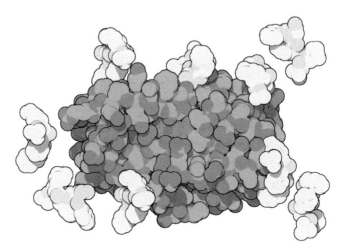

Figure 3-14 Drugs may be designed and tested in the computer. Automated docking techniques are used to find the best site for a drug to bind to the target bionanomachine. If the predicted binding is strong enough, the molecule may then be synthesized and tested for activity. The drug saquinivir is shown here binding to the HIV protease. Many conformations, shown in pink, are tested in the computer, and eventually the best conformation, shown in red, is found bound deep within the active site. Computer-aided drug design has been instrumental in the discovery of drugs to fight AIDS and drugs to fight many other diseases.

modes, using the energetic model to determine which one is the best. Current methods use simplifications to make this process feasible with available computer hardware. Most often, the protein target is treated as wholly or partially rigid, reducing the number of conformations that must be searched. In some cases, this can cause significant problems, for instance, when a protein closes around a small molecule when it binds. Current techniques are successful in about half of the cases using off-the-shelf techniques. The remaining systems require special attention to deal with any protein motion that may affect the results.

Popular methods include:

(1) *AutoDock (Scripps Research Institute).* A genetic algorithm is used to search conformations and an empirical free energy force field is used to evaluate energies.
(2) *Dock (UCSF).* A geometric matching algorithm is used to match ligand structures to a simplified representation of the binding site, and then more sophisticated energetic models are used on the best solutions. This method is very fast, allowing large databases of small molecules to be docked and evaluated.

Predicting the interactions between large biomolecules, especially between two proteins, is still a considerable challenge. No turnkey methods are currently available, but many laboratories are testing new methods. The problem is considerably more difficult than predicting the interaction of proteins with small molecules because of the large size of both of the molecules. Two general approaches are under study. In the first, the biomolecules are simplified by using a smoothed representation of the surface and chemical properties. Candidate complexes are then obtained by using fast methods to dock these simple representations and then evaluating the best candidates in greater detail. The second approach is a brute-force atomic simulation, made possible by the advances in search technologies such as genetic algorithms. These are just now being reported and are remarkably successful, showing great promise for future prediction of bound conformations as well as predicting the binding energy of the complex.

New Functionalities Are Developed with Computer-Assisted Molecular Design

Many of the recent successes in bionanotechnology involve the design of new functionality into natural biomolecules. The closer we stay to known structures, of course, the greater our confidence that the prediction will be realized in the biomolecule. The workhorse for design in many laboratories is simple molecular modeling, designing changes by hand on the computer graphics screen. These techniques are available off the shelf, and they allow researchers to build in new structure and then minimize the structure, looking at how well the modifications fit into the overall structure.

This approach, combined with molecular dynamics, has fueled much of the excitement in the molecular nanotechnology popularized by Drexler. A variety of nanoscale models have been built with a diamondoid lattice of atoms, using the appropriate bonding geometries. In a practical application, molecular modeling is used extensively for design of site-directed mutations for increasing stability and shifting functionality in proteins. This type of modeling requires creativity and experience on the part of the researcher, because, quite literally, any structure imaginable can be modeled and optimized.

Many laboratories have developed specialized methods to aid in molecular design. For instance, many methods have been developed to remove the manual effort from the design approach. These automated methods allow comprehensive searches instead of the intuitive hit-or-miss approach of modeling by hand, and tighter restraints on the energetics.

Many of these techniques are being developed in the field of computer-aided drug design. The goal is to design a drug molecule that perfectly fits into the active site of a target bionanomachine, blocking normal function. Some approaches begin by docking thousands of small fragments, each composed of 5 or 10 atoms. The best fragments are then linked to fragments that bind in neighboring portions of the active site, to form larger drug molecules. Another approach starts with a "seed" molecule that binds in the middle of the active site. The drug is then grown into a larger molecule by adding atoms one at a time until the active site is totally filled. These methods can design excellent candidates for new drugs but may occasionally run into problems with overeager researchers who design exotic molecules that are impossible for any chemist to build.

STRUCTURAL PRINCIPLES
OF BIONANOTECHNOLOGY

4

*At the atomic level, we have new kinds of forces and new
kinds of possibilities, new kinds of effects.*
— Richard Feynman

Our first goal in nanotechnology is to build a stable nanostructure. Only
then, after we can arrange atoms the way we want, can we can start think-
ing about what jobs these structures might do. To achieve this basic goal,
we must understand the forces that link atoms together inside a nanostructure. These forces are different than anything in our familiar world.
First of all, we can't shape atoms into any arbitrary form. When building
macroscale machines, we mold plastic, glass, or metal into any desired
shape. But at the nanoscale, our building material places more strict limits
on the possible shapes. Atoms bond to one another through a defined set
of chemical rules, and the shape of our nanomachinery will be restricted
by these rules. We also must be careful of the stability of our nanoscale objects: We must engineer them to be stable enough to withstand the particular environmental conditions under which they will be operating. The
stability is also limited by the small set of ways that atoms may be connected.

Fortunately, scientists have been laying the groundwork for this goal
for centuries. Chemists have discovered a wealth of information on the
structure and stability of molecules and have perfected methods for con-

Bionanotechnology: Lessons from Nature. By David S. Goodsell
ISBN 0-471-41719-X Copyright © 2004 John Wiley & Sons, Inc.

structing them from their component atoms. Biologists, on the other hand, have studied the atomic details of thousands of working nanomachines, each constructed by using only the basic principles discovered by chemistry. In this chapter, we will look at the different ways that bionanomachines are built. This is priceless information. We can analyze the ways that existing bionanomachines achieve stability and function and then use this information to develop a basic nanoscale toolkit for designing and constructing our own nanomachinery.

NATURAL BIONANOMACHINERY IS DESIGNED FOR A SPECIFIC ENVIRONMENT

When using biology as a guide for nanotechnology, we must keep an important limitation in mind. Natural bionanomachines are made to function inside cells. They have been optimized for this environment and may not function optimally, or even at all, when placed in different environments.

The most important limitation is the need for water. Bionanomachines are designed to be stable when surrounded by water. The unusual properties of water, described below, are harnessed to stabilize biomolecular structures. Except in rare cases, bionanomachines cannot be designed or analyzed in other solvents or in vacuum, because they only show their true structure and function when placed in water.

The biological environment is also limited to a narrow range of temperatures. Typical bionanomachines perform best at temperatures of about 37°C, although in special cases biomolecules may be designed to perform at temperatures up to 90°C (see below). Bionanomachines are designed to be stable at this temperature, but not too stable. At typical body temperature, thermal energy is manifested as a constant motion of molecules and the water surrounding them. The forces holding bionanomachines together are strong enough to build a stable structure despite the constant jostling of thermal motion and battering by water molecules. However, the forces are weak enough to allow the construction and demolition of bionanomachines with modest energy resources. Cells do not use arc welders and blast furnaces to forge new structures, instead, they perform all their synthetic and housekeeping tasks with the minimum expenditure of energy.

Natural bionanomachines are also constructed to be stable over a typical biological time scale. Most bionanomachines are expected to be functional for mere seconds, and bionanomachines are only rarely built to last more than a year. These machines are built quickly, used for a specific task, and then demolished, providing raw materials for building the next machine. Planned obsolescence is the rule.

Organic molecules based on carbon are ideal for building machines with these properties. Organic molecules provide a rich palette of interactions within the watery environment. They are stable at biological temperatures, but not too stable, allowing rapid synthesis or breakdown in a matter of seconds. By combining carbon with a few other types of atoms—oxygen, hydrogen, nitrogen, sulfur, phosphorus—a boundless variety of molecules, with diverse chemical properties, may be designed.

A HIERARCHICAL STRATEGY ALLOWS CONSTRUCTION OF NANOMACHINES

The dream of molecular nanotechnology is to build a nanostructure one atom at a time, starting from one corner and continuing atom by atom until the structure is finished. However, the approach taken by both chemists and nature is different. It is hierarchical, building large structures in several steps. George Whitesides has broken this hierarchy into four strategies for the construction of nanostructures, which build from the lowest level of atoms to the highest level of complex assemblies (Figure 4-1).

The first strategy is *sequential covalent synthesis*. Atoms are directly bonded into covalent molecules of the desired shape. Of the four hierarchical strategies, this one is the most similar to manufacturing techniques in our macroscale world. The product is designed, and then the components (atoms) are placed together piece by piece to build up the structure. This is exactly what synthetic chemists do. Synthesis of molecules such as vitamin B_{12} and taxol, with up to several hundred atoms, shows the upper limits of molecules that are currently feasible by synthetic chemistry. The advantages of covalent synthesis lie in the diversity that is achievable. Atoms may be combined in nearly any combination, including highly strained shapes and unlikely combinations of atoms, given, of course, that

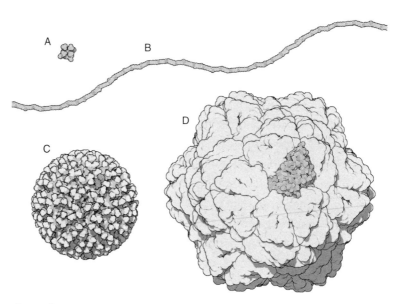

Figure 4-1   Several hierarchical strategies for construction may be used for bionan-otechnology. A. Covalent synthesis has been used to create many small organic molecules that mimic biological function. The molecule shown here is a mimic of the heme group in hemoglobin and shows similar binding properties for oxygen. B. Covalent polymerization has been used to create novel structural materials, ranging from tough plastics to elastic rubbers. A nylon chain, composed of small repeating units, is shown here. C. Self-organizing synthesis has been used to create liposomes for delivery of drugs. The lipids associate randomly to form a dynamic aggregate. D. Self-assembly is used to build the most complex biological machinery. The viral capsid shown here is comprised of 60 identical subunits arranged in perfect icosahedral symmetry.

the methods for proper positioning and bonding of the atoms are available.

The second strategy is *covalent polymerization*. Structures are built of modular units, which are linked into linear or branched chains. The synthesis may be performed in bulk, to form a mixture of chains. Plastics such as polyethylene are an example. Or polymerization may be controlled one step at a time, creating identical chains each time. The chemical synthesis of DNA by solid-phase techniques and the synthesis of DNA in cells are examples—in both cases, exactly the same chain, identical at the atomic level, is produced each time. Extremely large covalent molecules may be constructed by covalent polymerization, but it has inherent limitations. First, once the chemical schemes for attaching individual monomers is chosen, the chains are limited to that type of linkage. For instance, proteins will always be composed of a chain of linked peptide groups, not anhydrides or esters. Second, synthesis is limited to monomers that are stable under the reaction conditions, so some useful chemical groups may be too labile for use. Elaborate chemical schemes for protecting sensitive groups during synthesis have been developed to help solve this problem. In biological systems, enzymes allow much milder conditions to be used for polymerization, allowing use of monomers with a wider range of chemical properties.

The third strategy is *self-organizing synthesis*. Modular units are again applied, but the nanostructures are formed by noncovalent association of units. Familiar examples include molecular crystals, such as sugar crystals or protein crystals, and liquid crystals used in computer displays. In cells, examples include the micelles and bilayers formed by lipids. Many current applications termed "nanotechnology" fall under this category, such as nanospheres and nanocomposites. Note the difference between self-organization and the previous two levels of the hierarchy. In covalent synthesis and polymerization, the engineer links atoms together in any desired conformation, which doesn't necessarily have to be the energetically most favored position. Self-organizing molecules, on the other hand, adopt a structure at a thermodynamic minimum, finding the best combination of interactions between subunits but not forming covalent bonds between them. In self-organized structures, the engineer must predict this minimum, not merely place the atoms in the location desired.

The fourth strategy is *self-assembly*. Whitesides defines self-assembly as "the spontaneous assembly of molecules into structured, stable, noncovalently joined aggregates." I will include two processes in self-assembly. The first (which strains Whitesides's definition) is *protein folding*: the spontaneous folding of a protein chain into a stable, globular structure. The second is the classic conception of self-assembly: the assembly of globular subunits into defined multichain complexes. Both processes involve searching of many possible conformations until the thermodynamic minimum is found, powered by random thermal fluctuations. Highly specific interactions define the geometry of the final structure. Of all the lessons that may be learned from nature, the use of spontaneous self-assembly to construct nanomachines is arguably the most important. This mode of construction is utterly foreign to our macroscale technologies. Familiar manufacturing is dominated by willful, directed construction of objects based on specific blueprints that specify the three-dimensional form of the product. Cells, on the other hand, bring all of the necessary components together and let them self-assemble into the product.

THE RAW MATERIALS: BIOMOLECULAR STRUCTURE AND STABILITY

Carbon is the key to bionanotechnology. Organic molecules, built around carbon, are an ideal raw material, providing a wide range of design options for the construction of bionanomachinery. The diverse, stable bonding modes of carbon allow the construction of nearly any geometry that one might imagine. Upon this carbon scaffolding, atoms like oxygen and nitrogen may be added to incorporate additional molecular properties and functionalities.

The structure and properties of organic molecules may be understood by using a simple empirical description, tried and tested over the years by chemists and biologists. This is not a full description, such as that provided by quantum mechanics, but it is sufficient for understanding the primary forces that shape and stabilize bionanomachines. This simple description includes three basic concepts. First, covalent bonding connects atoms to one another in stable, defined geometry. Second, several types of nonbonded forces control the interactions within molecules and between molecules. Finally, the emergent properties of water strongly modify the form and stability of molecules (Figure 4-2; Table 4-1).

Molecules are Composed of Atoms Linked by Covalent Bonds

The strongest interactions within biological molecules are covalent bonds formed directly between two atoms because of the quantum mechanical sharing of electrons. Covalent bonds are quite stable at biological temperatures, and a significant amount of energy must be spent to create and break them. The strength of most organic materials is a consequence of covalent

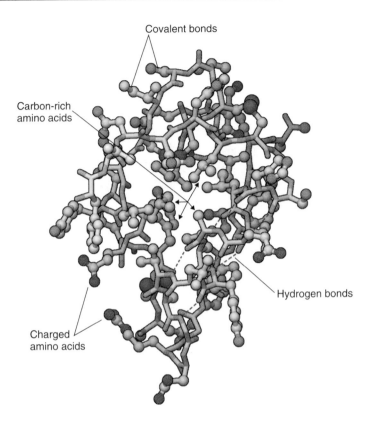

Figure 4-2 Insulin is a small protein stabilized by a collection of different forces. Covalent bonds, shown here as cylinders, connect the atoms of the structure. When placed in water, these chains fold to shield carbon-rich portions of the chain inside and to display the charged amino acids on the surface, where they can interact with water molecules. Hydrogen bonds connect different regions of the chain, strengthening the structure. The red dotted lines show hydrogen bonds formed in an α-helix.

bonding. For instance, silk is formed of many long strands of covalently bonded atoms, arranged side by side. Silk is quite resistant to stretching because of the strength of these bonds. It is flexible, however, because the strands slide next to one another freely, because neighboring strands are not connected by covalent bonds. If covalent bonds are created in all three dimensions, such as in the lattice of carbon atoms in diamond, the hardest known solids are formed, which are resistant to forces in all directions.

Covalent bonds are stiff and highly directional. The preferred geometry of covalent interactions may be understood through study of the preferred quantum mechanical states of the electrons. A full description of quantum

Table 4-1   Strength of Forces Stabilizing Proteins

	Strength (kcal/mol)
Covalent bonds	>50
Dispersion forces	<1
Hydrogen bond	1–7
Electrostatic interaction (low dielectric)	1–6
Hydrophobic interaction (two phenylalanine side chains)	2–3
Average thermal energy (37°C)	0.6

Source: Adapted from Devlin, T.M. (1992). *Textbook of Biochemistry with Clinical Correlations*. Wiley-Liss, New York.

mechanics is beyond the scope of this book, but, fortunately, a simple set of empirical rules is enough to understand most of the organic molecules that will be encountered in bionanotechnology. It is almost as simple as building with atomic Tinkertoys, using a standard set of atomic parts. However, for a deeper understanding, and for incorporation of more exotic atom types, a more detailed study of quantum mechanics and chemistry is necessary.

To a first approximation, organic molecules may be designed by snapping together atoms using the appropriate number and geometry of bonds, as shown in Figure 4.3. Using these simple rules, a diverse set of molecules may be designed. The rules are so general that plastic models are available for studying and designing organic molecules. However, when creating a new molecule, a little chemical intuition is necessary. Careful examination of the molecules that are actually built by cells and chemists will provide an idea of which molecules will be stable when actually constructed. In general, these molecules will have carbon skeletons, with oxygen and nitrogen atoms separated and not directly bonded to one another. Also, molecules that distort the normal geometry of the atom, such as a tight triangular ring of three carbon atoms, will probably be less stable and thus more difficult to construct.

The geometry of interaction through covalent bonds is well defined and relatively rigid. The lengths of bonds vary by only a fraction of an angstrom, and flexing of the angles formed between two bonds rarely exceeds a few degrees. Rotation around bonds, however, is common for many bond types. In general, single bonds are surprisingly flexible and have been proposed for use as rotatable axles in molecular nanotechnology. However,

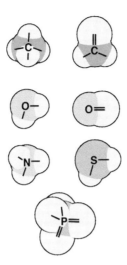

Figure 4-3   Organic molecules may be constructed with a few simple geometric relationships between atoms. Carbon atoms (C) make four bonds with surrounding atoms, in tetrahedral arrangement. In some compounds, two (or even three) of these bonds may form with a single atom, forming a double bond. Oxygen (O) forms two bonds in a bent geometry or a double bond to one neighboring atom. Nitrogen (N) forms three bonds, and sulfur (S) forms two. Phosphorus (P) is encountered as the phosphate group in bionanomachinery, with the phosphorus atom surrounded by four oxygen atoms.

double and triple bonds, where additional electrons are shared through bonding, are rigid.

As with most things in life, the story is not quite this simple. These simple rules must be amended with the concept of *resonance* to account for some unusual anomalies. Benzene is the classic example. Our rules would have us draw a structure with three single carbon-carbon bonds alternating around the hexagonal ring with three double carbon-carbon bonds. As you can see, there are two possible ways to create this structure. But when the molecule is actually studied, it is found that all the bonds are exactly equal. In reality, the structure is intermediate between these two extremes.

Resonance has a profound effect on the flexibility of the molecular structure. The peptide bond used extensively in bionanomachinery is an example of this. One would predict that the central single bond would be flexible, allowing rotation around the middle. In fact, this is not observed—the whole peptide group acts as a single rigid unit. This may be understood by looking at an extreme form of the peptide unit, with an electron transferred from the nitrogen atom to the oxygen (as shown in Figure 4.4). It has a double bond across the middle, holding the structure rigid. In reality, the electronic structure is again intermediate, but enough like the double bond to hold the entire group rigid.

Dispersion and Repulsion Forces Act at Close Range

The overall shape and form of molecules is a consequence of two opposing forces, which define the inviolable space of each atom. As atoms approach one another, these two opposing forces define the interaction (Figure 4-5).

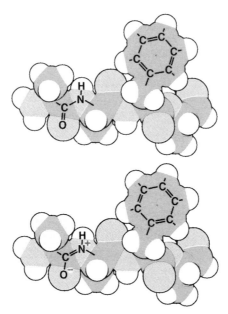

Figure 4-4   Resonance rigidifies parts of bionanomachinery. The peptide bond used to construct proteins is rigid because of resonance. In the upper structure, the peptide bond shows the typical bonded structure, which would be flexible around the C–N single bond, and the lower structure shows an alternate form, which would rigid around the C=N double bond. In reality, the bond is intermediate and the actual structure is rigid, resisting rotation around the C–N bond. Some rings of atoms also show resonance, which leads to a rigid, flat structure. The ring of six carbon atoms shown here can be thought of as a mixture of two structures, with different placement of C–C single and double bonds. This type of resonance also stabilizes the flat rings of DNA bases.

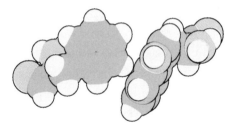

Figure 4-5 Dispersion-repulsion forces act at close range. Here, two large amino acids are touching, held loosely together by dispersion forces but kept from overlapping by strong repulsion forces. These forces are conveniently visualized by using "spacefilling" models, like those used for nearly every illustration in this book. These models show each atom as a sphere appropriately sized to represent the position of the most favorable balance of dispersion and repulsion interactions.

When atoms are close, *dispersion* forces provide a weakly favorable interaction, drawing the atoms together. But if the electron clouds begin to overlap, strong *repulsion* forces dominate, keeping the atoms from interpenetrating. Together, dispersion and repulsion forces form a highly unfavorable wall preventing overlap of atoms and a weakly favorable well coaxing atoms to remain in contact, just touching one another. Dispersion-repulsion interactions add significantly to the stability of bionanomachines. Each interaction is energetically small, but these forces quickly add up for the many atoms in bionanomachines.

Repulsion forces are a consequence of the Pauli exclusion principle, which states that two electrons cannot occupy the same quantum mechanical state. As neighboring atoms approach one another, this force rises rapidly to very unfavorable energies as the electron clouds overlap. Repulsion forces are highly unfavorable at short distances, but once the atoms are no longer in contact they fall to negligible values. These forces dominate when two objects collide, so that billiard balls ricochet instead of passing through one another. Repulsion forces also play an indispensable role at the molecular level, by limiting the conformational freedom of bionanomachinery, as described below in the section on flexibility.

Dispersion forces counter these repulsion forces when atoms are just touching. Dispersion forces are caused by induced charge interactions. When the negatively charged cloud of electrons is close to a neighboring atom, the nearby electrons in the neighbor induce a small dipole in the atom, by shifting some of its electron cloud slightly away from the neighbor and slightly exposing the positively charged nucleus. This transient dipole interacts favorably with the neighboring electron cloud, forming a weakly attractive force. Dispersion effects are felt when two surfaces come into intimate contact, contributing to the macroscopic properties of adhesion and friction. For instance, flies and geckos use dispersion forces between their feet and a surface to climb vertical walls. The surfaces are pressed tightly together, so dispersion forces are strong enough to adhere them to the surface.

Hydrogen Bonds Provide Stability and Specificity

Hydrogen bonds play a central role in the stability of bionanomachines, and in their interactions with one another. Hydrogen bonds are like Velcro; they are reusable fasteners that may be connected and broken according to need. Hydrogen bonds are weaker and less directional than covalent bonds, but they are slightly stronger than the typical thermal energy, so they are stable in biological contexts. Because they are not as stable as covalent bonds, they are not as difficult to break. The surface tension of water, causing water to bead into tight drops, is a physical manifestation of the strong hydrogen-bonding force between water molecules, but the liquidity of water is a manifestation of the ease of making and breaking hydrogen bonds at room temperature.

Hydrogen bonds are thought to be mainly electrostatic in nature. They are formed between (1) a hydrogen atom that is bonded to oxygen, nitrogen, or sulfur and (2) another oxygen, nitrogen, or sulfur atom. The interaction is somewhat directional, forming the strongest interaction when the hydrogen atom is aligned directly at the accepting atom (Figure 4-6). Nonoptimal hydrogen-bonding geometries, however, are very common in natural bionanomachines.

The abundance of oxygen and nitrogen atoms in biological molecules makes hydrogen bonding a significant source of structural stability. Hydrogen bonds are widely used within bionanomachines to stabilize internal structure. They also play an essential role in molecular recognition, as described in Chapter 5. In addition, the water environment of bionanoma-

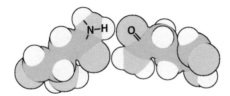

Figure 4-6   Hydrogen bonds are directional. They are strongest when the hydrogen atom on the donating group, shown here on the left, points directly at the accepting group, shown here on the right.

chines gives hydrogen bonding additional importance. Each water molecule displays two sites for accepting hydrogen bonds and two hydrogen atoms to participate in a bond. The surfaces of bionanomachines are continually exposed to water, so the ability to form hydrogen bonds with water is essential for biomolecular stability. This important topic is described in more detail below.

Electrostatic Interactions Are Formed Between Charged Atoms

Classic electrostatic interactions between atoms with electronic charges also play an important role in the function and stability of bionanomachinery. Electrostatic forces are long-range forces, extending over many times the diameter of a single atom (Figure 4-7). Electrostatic forces are used both at short distances, to form a strong linkage between groups, and at longer range, to attract or repel other molecules that may be encountered. Electrostatic forces act between charges on atoms—opposite charges attract, and like charges repel one another. These forces are nondirectional, acting symmetrically in all directions from each charge center.

Several types of electrostatic interaction are common in bionanomachines. The first is a *salt bridge* formed between two organic groups that carry a formal charge (the name indicates the similarity between these charge-charge interactions and the interactions between ions in inorganic salt crystals). These are common on the surface of natural biomolecules, where they serve to stabilize the structure. When extra strength is needed, biomolecules will incorporate charged sulfate and phosphate groups.

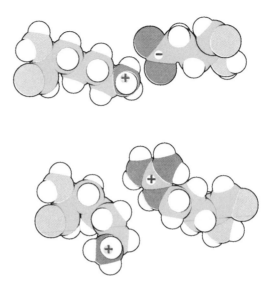

Figure 4-7 Electrostatic interactions act over long ranges. Groups with opposite charges, like the lysine and glutamate amino acids at the top, attract one another. This forms a strong salt bridge that is often used to glue different parts of bionanomachines together. Similar charges, like the positive charges on lysine and arginine shown at the bottom, repel one another. This can be used to inhibit interaction of two molecules.

Second, bionanomachines often incorporate charged metal ions to stabilize a structure or for use in specific chemical functions (Figure 4-8). These ions range in size from light magnesium ions to heavy atoms of iron and cobalt. Often, the ions must be trapped within special chemical containers that coordinate the ion in the proper orientation and maintain it in the appropriate electronic state.

Electrostatic interactions are reduced by the *dielectric effect*, which is dependent on the atoms between and around the charged atoms. Water acts as a strong dielectric, strongly reducing interaction of ions, whereas protein is a weak dielectric. Water molecules are dipoles: The oxygen atom is somewhat negative, and the two hydrogen atoms are somewhat positive. When charged ions are placed in water solution, the surrounding water molecules tend to align their dipole moments to interact with the ion. This reduces the force on other ions, dampening electrostatic interactions by 80-fold. The atoms in proteins, on the other hand, are more fixed and cannot undergo this dynamic reorganization.

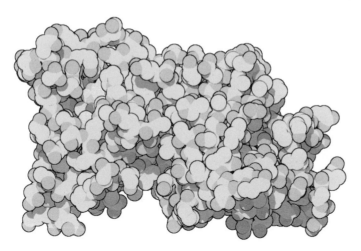

Figure 4-8 Ions are widely used in proteins to perform specific chemical tasks. The protein superoxide dismutase, which detoxifies reactive forms of oxygen, is shown here. The active site uses a copper ion and a zinc ion, shown in red, to bind and chemically modify superoxide. The metal ions are held within the protein by a collection of histidine amino acids, shown in pink.

The Hydrophobic Effect Stabilizes Biomolecules in Water

Water is a unique medium, creating a complex environment for the function of bionanomachines. The forces described above are easily understood by looking at molecules atom by atom: You can simply sum up the contributions of covalent bonds, hydrogen bonds, dispersion/repulsion forces, and electrostatics to predict how a bionanomachine might behave. When placed in water, however, the picture is far more complex. A surprising emergent property of water, termed the *hydrophobic effect*, dominates the properties and interactions of bionanomachines.

Water interacts strongly with itself, through its abundant hydrogen bonds. Liquid water is composed of shifting water molecules, continually forming and reforming hydrogen bonds with their neighbors. The stability of a water solution is a combination of enthalpic energies, such as these dispersion/repulsion and hydrogen bonds between the molecules, and entropic energies, which tend to favor states of the solution in which the individual molecules are in more random orientations. The hydrogen bonds between water molecules are enthalpically very favorable, because they form many stabilizing interactions. They are also entropically very favorable, because each water molecule has countless options for interacting with neighbors, all of essentially equal energy. Anything that interrupts this process must provide an equally favorable interaction with as many entropic options or it will be energetically disfavored.

However, bionanomachines are built predominantly of carbon, which interacts only weakly with the surrounding water. When a hydrocarbon molecule is suspended in water, the water molecules surrounding the hydrocarbon lose their ability to form hydrogen bonds freely with neighboring water molecules. On one side, they merely contact the hydrocarbon, forming a favorable dispersion interaction but losing the ability to form a stronger hydrogen bond. On the other side, they attempt to maximize the remaining interaction with the water solution, limiting their possibilities relative to a water molecule free in solution. Thus these water molecules that are pressed close to the hydrocarbon are unfavorable both in enthalpy and entropy.

Now, if we take a number of these hydrocarbon molecules and group

---

Lessons from Nature

- Organic molecules are excellent building materials for bionanomachines, providing:

  (1) defined structure and geometry through covalent bonding, and
  (2) multiple modes of nonbonded interaction with a wide range of energetic strengths.

- Dispersion/repulsion forces define the space-occupying properties of molecules.
- Hydrogen bonds and electrostatic interactions provide specificity and stability.
- The hydrophobic effect, an emergent property of water, stabilizes compact, aggregated forms of carbon-rich molecules in water solutions.

them together in one place, the situation improves. The total hydrocarbon surface that is exposed to water is reduced when they all associate together. Many water molecules are freed to return to the random, shifting solution. The hydrocarbons are energetically happy as well, as they interact with neighboring hydrocarbons inside the aggregate through their dispersion forces. This is the hydrophobic effect in action. Hydrocarbons aggregate together in water solutions, finding the arrangement that frees the most water. The hydrophobic effect is visible at our macroscopic scale in salad dressing: Oil and water do not mix. At the molecular scale, a similar separation of phases drives much of the process of self-assembly of bionanomachinery.

It is often convenient to think of this in terms of a "hydrophobic bond" stabilizing the association of carbon-rich molecules. One must remember, however, that this stabilizing interaction is a consequence of the freeing of water molecules and is not due to any intrinsic interaction between the carbon-rich molecules.

Far from being a detriment, the hydrophobic effect is exploited by bionanomachinery. Hydrophobic components play a role in nearly every type of bionanomachine. They drive protein folding and self-assembly (Figure 4-9). Proteins are constructed with carbon-rich amino acids that are sequestered inside when the protein chain folds into its stable structure. DNA and RNA strands associate into compact double helices to hide the flat, hydrophobic faces of their bases. Lipids associate into large double-sided membranes that hide their carbon-rich tails. Hydrophobicity drives the assembly of these machines and stabilizes the final structure.

PROTEIN FOLDING

Nature uses an information-efficient method for construction of proteins. Natural proteins are designed to form stable globular structures. The amino acid sequence of each protein is tailored to provide a collection of carbon-rich amino acids to form a stable core, taking advantage of the hydrophobic effect. They also include many charged and hydrogen-bonding amino acids, which tie the chain together into a stable bundle. Everything is in precisely the right place. But perhaps more remarkably, proteins are designed

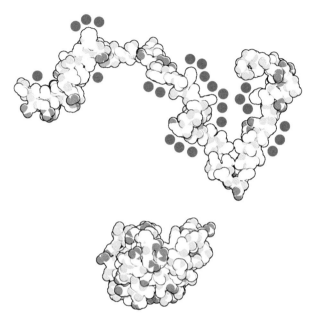

Figure 4-9   The hydrophobic effect drives the folding of proteins. The unfolded chain, shown at the top, has many regions that are rich in carbon, shown in white here. Water molecules, shown schematically as gray circles, form an unfavorable interaction with these regions. When the protein folds, as shown at the bottom, these water molecules are released to form more favorable interactions with the surrounding water. Note that the folded protein, lysozyme, is covered with nitrogen and oxygen atoms that carry electrical charges, shown in bright red, which form strong interactions with the surrounding water.

to fold spontaneously from a random conformation into this perfectly designed globular structure. The specifications for both the stable structure and the process of assembly are encoded in the sequence of amino acids. This is an amazing feat of design. A remarkably small amount of information can be used to construct highly complex machinery.

   The ability to predict the folded structure of a protein, given only its sequence of amino acids, is a major hurdle currently facing bionanotechnology. Once this problem is solved, it will immeasurably aid progress in bionanotechnology. Then we will be able to design new proteins with custom conformations and functions. The mechanism of protein folding is an area of active research, and many exciting questions have been answered.

Not All Protein Sequences Adopt Stable Structures

By current estimates, only a small fraction of the total number of possible amino acid sequences will fold to form stable structures. Researchers estimate that there are about 1000 ways to fold a protein chain to form a stable structure. However, this does not mean that there are only 1000 protein sequences that fold into stable structures. A wide range of similar, homologous sequences fold into similar structures, differing only in local details.

The variability of protein sequences was vividly demonstrated in a discussion by Max Perutz (Novartis Foundation Symposium 213, 1998). He noted that the sequence of hemoglobin is highly variable when comparing the forms made by different animals. Of the 140–150 amino acids in the hemoglobin chain, only two are conserved across the entire set of species: a histidine that directly coordinates with the essential iron and a phenylalanine that is essential for proper placement of the heme cofactor. All other amino acids may change to modify function in different ways (affinities for oxygen may differ by 100,000 times between different species) or simply through genetic drift. Despite this large range of sequence variability, all adopt a nearly superimposable folding pattern.

Structure prediction is quite robust for homologous sequences. For sequences with about 30–40% identical amino acids, there is a high probability that the folded structure will be similar enough to allow accurate modeling with current methods (see Chapter 3). This is a boon for bionanotechnology, allowing prediction of many new protein structures. The protein engineer, however, is quickly faced with a significant problem. These statistics apply for natural biomolecules, which have been selected by evolution for proper folding. A single change in the wrong place may be fatal but will still show essentially the same high homology to a protein that folds successfully. Thus, when modifying an existing protein for a new function, the modifications must be taken in small steps, ensuring that each retains the ability to fold into the desired structure.

Globular Proteins Have a Hierarchical Structure

Natural proteins have several levels of structure that appear to be important both in proper folding and in proper final structure and stability (Figure

4-10). Local regions of the chain adopt a few particularly stable structures, such as α-helices, β-sheets, and a variety of stable loops and turns. These structures maximize the number of hydrogen bonds between the peptide groups in the protein chain, forming stable interactions and also ensuring that these peptides will not be buried inside the protein without having the proper hydrogen-bonding partners.

These stable local structures then fold into a stable globular structure. α-Helices and β-strands are typically arranged into groups, which then associate into larger structural domains. Individual α-helices rarely interact directly with individual β-strands, because of a mismatch in the arrangement of hydrogen bonds. β-Strands associate side by side to align hydrogen

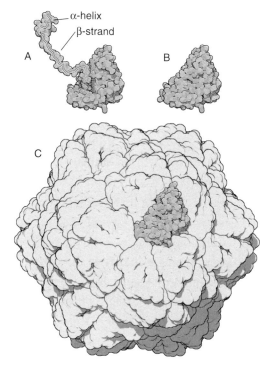

Figure 4-10   Most proteins are built with a hierarchical structure. A. The protein chain adopts several particularly stable local structures, such as the α-helix and the β-strand. B. These then fold to form a compact globule. C. Several globular proteins may then combine into an oligomeric assembly, such as the virus structure shown here, which has 60 identical subunits assembled into a hollow capsid.

bonds between strands, forming two-stranded ribbons, multistranded sheets, or cylindrical barrels. The specific geometry of the protein backbone favors a twisted conformation for these structures that is observed in nearly every natural protein. α-Helices satisfy all of their hydrogen bonds internally, leaving all of the side chains facing outward from the cylindrical core. The helical arrangement of side chains forms grooves and ridges along the cylinder. α-Helices tend to pack side by side to align their ridges into the grooves. α-Helices also commonly pack against the flat faces of β-sheets.

As a result of these favored interactions between α-helices and β-sheets, and the need to fold spontaneously from an extended chain to a compact globule, many proteins fall into a few similar folding patterns. A few of the most common are shown in Figure 4.11. Note that most are folded into simple topologies. Some are obtained by folding the chain in half and then in half again one or more times. Others are arranged like helical springs. Knots, with the chain threading through a looped portion, and other complex topologies are very rare. Jane Richardson systematized these folding patterns, noting their similarities, and highlighted their relationship to the process of protein folding. As more structures have been solved, new folding patterns are occasionally discovered, some surprisingly beautiful. However, most proteins fall neatly into existing stable folding patterns.

Finally, many globular structures may associate into a larger complex. In some cases, several different protein chains combine to form the complex. In other cases, as described below, symmetry is used to combine many identical subunits into a larger assembly.

Stable Globular Structure Requires a Combination of Design Strategies

The first requirement for protein folding may be thought of as *positive design*. A protein must designed to be energetically stable. Protein chains are designed to fold in water. The driving force for folding is the incorporation of carbon-rich hydrophobic amino acids that favor conformations that are shielded from water. Most stable protein structures have a *hydrophobic core*, a closely packed collection of carbon-rich side chains at the center of the protein. The side chains in this core, interacting through dispersion/re-

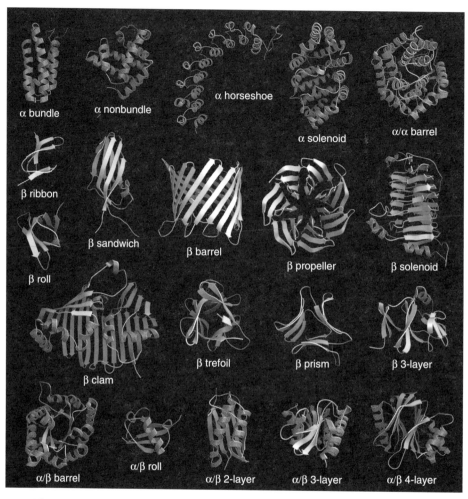

Figure 4-11 The combination of α-helices, β-sheets, and connecting loops allow the construction of many protein topologies. Some of the more common are shown here. A ribbon is drawn that follows the protein chain, with α-helices in pink and β-strands in flat white arrows.

pulsion interactions, often fit together like pieces in a puzzle, completely filling the available space and leaving no pockets large enough to enclose a water molecule. These interactions provide most of the energetic stabilization for folded globular protein structure. For instance, each leucine or phenylalanine amino acid will provide a favorable stabilization of about 2–5 kcal/mol, relative to an alanine. But these interactions are not sufficiently directional to specify a unique folded conformation of the protein chain.

Unfolded protein chains are highly flexible, so as the chain folds into a compact structure, the entropy is reduced. This is energetically unfavorable in an amount of about 1–2 kcal/mol per amino acid in the chain. To form a stable globular structure, this unfavorable reduction of entropy must be compensated by favorable interactions in the folded structure. When these hundreds of favorable interactions are added to the unfavorable entropy of folding of hundreds of amino acids, typical proteins are stabilized by a margin of about 4–10 kcal/mol. This energy of stabilization, although not overwhelming, is sufficient to favor the folded state by thousands to millions of times over the unfolded states.

A second essential requirement for protein folding is often termed *negative design*. The goal of negative design is to ensure that a single folded conformation is created. The protein chain is designed so that all unwanted conformations are energetically unfavorable, so that only the desired conformation is formed. As articulated by DeGrado and co-authors, "negative design is a critical process, which requires one to anticipate and destabilize as many possible alternative low-energy structures as possible."

Many negative design approaches are used in natural proteins. Polar and charged amino acids, in particular, are essential for negative design. They are placed on loops and other segments, forcing them to remain on the protein surface when the protein folds. Hydrogen bonds and salt bridges that are inside proteins are also used for negative design. In properly folded proteins the partners form stabilizing pairs, but when improperly folded the partners would be unpaired in the carbon-rich environment inside the protein. Pairs of charges are placed on the surface of the protein, which form stabilizing interactions in the proper fold but which would form strong repulsions in inappropriate folds. These uses of polar atoms are combined with shape-based approaches. Glycine and proline are widely used to interrupt the formation of α-helices, so they are common in regions that form loops in the proper structure. Finally, the different carbon-rich amino acids are used to build individual elements that fit together like a jigsaw puzzle in the proper structure but which would bump and clash in improper folds.

Chaperones Provide the Optimal Environment for Folding

Many biomolecules require some assistance when folding and maturing. Cells contain a collection of biomolecules, collectively known as *chaperones*, that assist in folding, as well as a wide variety of biomolecules that process the proteins into mature forms (Figure 4-12). This is important to keep in mind when engineering an organism to create a protein. Different organisms have different systems for assisting in folding and maturation, so the end product may be different.

Stretches of hydrophobic amino acids are essential for protein folding, but they also present a potential danger. Protein folding is driven by the need to shelter these carbon-rich stretches from water. This may be accomplished, however, in the same way that lipids hide their hydrophobic regions—by aggregation into large membranes or amorphous clumps. To solve this potential problem, chaperonins separate the individual chains and provide an environment in which they can fold without interference. Chaperonins are canisters with a hydrophobic interior. New protein chains enter the interior, and the top is closed. This induces a large change in the chaperonin that rotates the walls, hiding much of the hydrophobic character and creating a larger space. The protein, losing the stabilization of the walls, is then forced to fold on its own.

Other proteins are built as large precursors, which undergo modification after proper folding. Insulin is a common example. It is a very small protein composed of two separate chains. Its stability relies in large part on several internal disulfide linkages. It would be difficult to design an insulin-sized protein that would fold spontaneously into a stable form. Instead, it is created as a much longer precursor and then the large extraneous loop is removed by cleavage to yield the mature protein.

Two processes greatly limit the speed by which proteins adopt their native structure. Naturally, cells have two chaperones that assist these processes. The first is the formation of disulfide bonds that cross-link cysteine amino acids at distant parts of the polypeptide chain. Improper formation of these linkages can lock the protein into an improper conformation. The enzyme *protein disulfide isomerase* assists with this process by breaking inappropriate linkages. The second problem is encountered with proline amino acids, which may adopt two different molecular conforma-

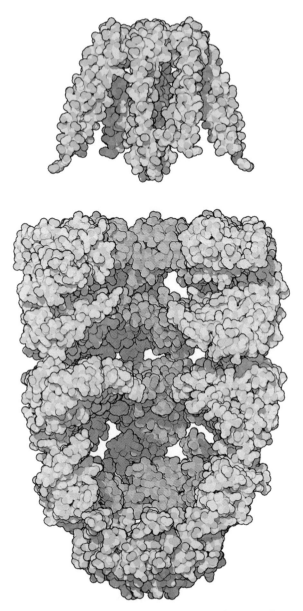

Figure 4-12   Chaperones assist the folding of proteins. Prefoldin, shown at the top, is shaped like a hand. It grabs new protein chains and keeps them from aggregating with other new chains. The bacterial GroEL chaperonin, shown at the bottom, guides protein folding. It is cylindrical, with two large spaces inside. The side facing upward is open in this picture. Protein chains enter this cavity and interact with hydrophobic amino acids on the walls. Then, when capped by GroES, shown in gray on the lower half, the cylinder expands and the hydrophobic amino acids are shielded, forcing the protein inside to fold on its own.

tions. Typically only one will have the appropriate structure for fitting into the tight spots often occupied by proline. The enzyme *prolyl peptide isomerase* assists in the interchange of these two conformations in folding proteins.

Note that chaperones are not specific templates or blueprints for assembly. Instead, they provide the proper environment for folding or they assist proteins over common protein folding hurdles. As such, they do not require information about the particular protein and they do not provide information used in the folding of the protein.

Rigidity Can Make Proteins More Stable at High Temperatures

Once proteins fold into globular structures, they are prone to unfolding. Extremes of temperature, salinity, pH, and other environmental factors promote unfolding and loss of activity in proteins. One of the earliest industrial applications of protein engineering was to stabilize enzymes for use in nonbiological environments. One goal of this work is to create enzymes that are stable at extreme temperatures. Fortunately, we can look to nature for clues as to how to proceed. Thermophilic bacteria, which live at temperatures from about 60°C to 80°C, and hyperthermophilic bacteria, which live at temperatures of 110°C and above, have developed proteins that are stable and highly active at these elevated temperatures. Their proteins are attractive for industrial applications because they are typically more stable to solvents and they allow reactions to be performed at high temperatures, allowing the use of higher concentrations of starting materials and often showing faster reaction rates. Their stability at high temperature also allows for easy purification: You can simply heat up an impure mixture and all of the contaminating proteins will be destroyed.

Surprisingly, the structures of these heat-stable proteins show only subtle differences from the heat-labile counterparts made by other organisms. The amino acid sequences of heat-stable proteins are typically 40–85% similar to their counterparts in other organisms and their three-dimensional structures, compared to those of other organisms, are nearly identical. They also typically use identical catalytic mechanisms. They are highly similar in structure and function but have somehow layered heat stability on top of the typical properties.

The major difference appears to be an increase in rigidity. The stabilization is not afforded by improvements in the packing of the hydrophobic core. The core residues are already packed with maximal efficiently in typical proteins, so there is little room for improvement in heat-stable proteins. Instead, heat stability appears to be afforded by small changes at the surface of the protein, adding small measures of stability and rigidity that resist the initial unfolding events that serve to inactivate the protein. These include new ion pair interactions between charged amino acids, new disulfide linkages or metal ions, incorporation of rigid proline amino acids, or replacement of flexible glycine amino acids. Often, these new interactions tie down a loop that is normally flexible or immobilize one of the ends of the protein that is usually free and flexible. Also important are changes that replace or protect amino acids that are destroyed by heat, such as glutamine and asparagine, which have sensitive amide groups.

Enzymes from thermophilic and hyperthermophilic organisms are optimized for function at high temperatures, from 70°C to 125°C. We would expect that the rates of the enzyme reactions would be much faster than rates in typical organisms, given that many chemical reactions double in speed if the temperature is raised by 10°C. Natural thermophilic enzymes, however, have reaction rates that are comparable to the rates of typical enzymes. There is a trade-off during the evolution of these enzymes: Catalytic efficiency is lost as heat stability is gained. Because there is no selection pressure to improve catalytic function above the levels found in typical organisms, thermophiles have not developed superenzymes that take full advantage of the high temperature. This goal is left to the protein engineers with industrial applications that can benefit from ever-faster reaction rates.

Many Proteins Make Use of Disorder

In most cases, we think of proteins as having a defined, stable structure. They may have a few moving parts connected by hinges or loops, but the overall structure is well-defined. However, it is becoming increasingly apparent, with the analysis of sequences of entire genomes, that a significant number of proteins do not adopt a single defined structure but instead utilize specialized transitions between disorder and order in their function.

For instance, the hormone glucagon, a small peptide of 29 amino acids, is unstructured in solution. But when it binds to its receptor, it adopts a specific conformation for recognition. Many signaling proteins become ordered only on binding to their proper partner. Order-disorder transitions are particularly attractive in signaling proteins. Disordered proteins are rapidly digested inside cells, so they will have a short functional life, allowing quick responses and tight control of signals. Also, disorder allows the same protein to be used in many capacities, allowing a single protein sequence to be recognized in several ways. Signaling pathways in cells are very complex, and some signaling proteins have many varied effects on different systems.

Other examples include molecules that bind to very large targets. DNA-binding proteins often employ long disordered linkers. For instance, p53 protein, which acts as a watchdog for cancer, is composed of four arms that bind to four neighboring sites on the DNA (Figure 4-13). The four chains are held together at the center by a packsaddle linkage, and the DNA-binding

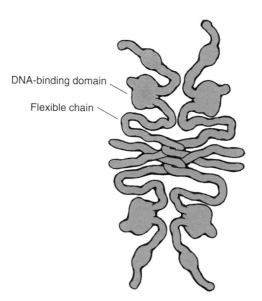

DNA-binding domain

Flexible chain

Figure 4-13 The p53 protein uses disorder to allow binding to four separate sites on a DNA strand. The protein contains four DNA-binding domains, tethered together by flexible protein chains.

Lessons from Nature

- Design of spontaneously folding globular proteins includes positive design, engineering a stable conformation of the chain, and negative design, engineering a chain that disfavors folding into improper conformations.
- Only a small fraction of the possible amino acid sequences spontaneously fold into stable structures, but many similar amino acid sequences fold into the same conformation.
- Proteins have a hierarchical structure, with local structure such as α-helices and β-sheets associating to form stable globular proteins, which are then assembled to form larger structures.
- Chaperones are often needed to inhibit aggregation of new protein chains and to assist the process of protein folding through structural bottlenecks.
- Selective rigidity of surface features can slow the unfolding of proteins, making them more stable at high temperatures.
- Selective disorder may be incorporated for specific functions by proper choice of amino acid sequence.

---

domains are tethered on the four arms. Antibodies are another example, with a flexible linker allowing free motion of the antigen-binding arms, allowing them to fit onto differently shaped targets.

Disordered proteins must be specially designed to discourage the formation of stable folded structures. This can be accomplished by including soluble amino acids and avoiding carbon-rich amino acids. The use of glycine and proline also favors disorder: Glycine is very flexible, and proline forms a kink.

SELF-ASSEMBLY

Imagine throwing a collection of fenders, engine parts, nuts and bolts, wheels, and axles into a swimming pool and expecting a functional car to drive out. As remarkable as it might seem, this is how most natural bio-nanomachines are built. All of the components are built as flexible chains, which spontaneously fold into compact structures and finally assemble spontaneously into functional complexes. This process has been termed *self-assembly*.

Self-assembly is used throughout biology. It is made necessary by the bottom-up assembly technique used in living cells. They specify the building of three-dimensional bionanomachines with only the one-dimensional information held in DNA. Protein chains are built with the information in DNA, and then they have to adopt their final structure without further input of information.

This is quite different from the directed assembly techniques used in macroscopic engineering. We are continually adding information during the entire assembly process. Each piece is carefully fabricated with a desired shape, and then each piece is individually guided into place. The speculative assembler-based approach of molecular nanotechnology is similar, specifically placing each atom in its proper place. Bionanomachines, on the other hand, show that this level of control is not always necessary to create a functional nanostructure.

Self-assembly is a revolutionary concept. Researchers in diverse disciplines, working at all scales, have embraced the concept and are applying it to different tasks. A few examples of the applications that they are pursuing are given in Chapter 6.

The design of self-assembled structures is far more restrictive than directed construction, where intelligence and external control may be added at each stage of assembly. Each part must be designed to form a stable complex, but each part must also encode the entire mechanism for assembling spontaneously into this complex and evading improper complexes. A few general design principles used in natural bionanomachines promote self-assembly.

Self-assembly is *modular*. Modularity provides several advantages. First, large assemblies may be created with many identical modules, like bricks in a wall. This allows a large assembled structure to be specified with a modest amount of information. In biological systems, identical subunits are often assembled by using symmetry to control the size of the final complex. Modularity also provides for error control. Individual modules that are constructed incorrectly may be excluded during the assembly.

Self-assembly requires *specific geometry of interaction* between subunits. Modules assemble in defined orientations, forming assemblies with defined geometry instead of random aggregates. This is accomplished through surfaces that match in shape and that associate with many weak, reversible interactions, as described more fully in the section on molecular recognition.

Self-assembly requires *unique interaction* between subunits. Within a given system, each interaction used to construct the assemblies must be unique and different enough from the other assemblies to exclude any cross talk. In a cell, this requires thousands of structurally unique interfaces. This can be a severe problem with small molecules, where the interaction surface is small. For instance, this problem plagues the design of therapeutic drugs. A small organic molecule designed for use as a drug will ideally target only a single protein, but often it also binds to other proteins with similar active sites, causing unwanted side-effects. This does not appear to be a significant problem with protein interactions, however, because interacting surfaces of proteins are typically much larger than the limited surfaces of interaction between proteins and small organic molecules.

Self-assembly is *spontaneous,* requiring no input of information to guide assembly. Self-assembly relies on thermodynamics to determine the final structure of the assembly. This involves a careful trade-off of enthalpy and entropy. The enthalpy of the many new dispersion and hydrogen-bonding interactions is generally favorable, whereas the entropy of locking many subunits into a single complex is unfavorable. Cooperativity, where binding of one subunit to another increases the affinity for additional subunits, often modifies the kinetics of this spontaneous association. This often leads to "all-or-nothing" construction, so that once started the entire complex quickly assembles.

Symmetry Allows Self-Assembly of Stable Complexes with Defined Size

Symmetry provides many advantages for the self-assembly of large complexes from modular subunits (Figure 4-14). The most obvious advantage is

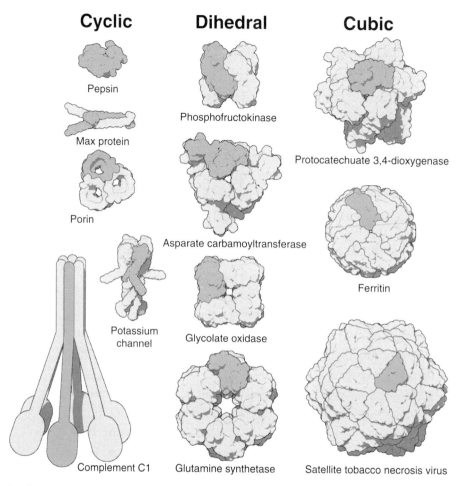

Figure 4-14 Symmetry is used for many functions in natural biomolecules. Examples of proteins with the most common point group symmetries are shown here. The first column shows examples with cyclic symmetry, in which subunits are arranged around a single axis of rotation. The simplest is a simple monomer with no symmetry. This is commonly found in simple enzymes like pepsin. Twofold symmetry is used by Max protein to create calipers that grip DNA. Two channels take different approaches to forming pores through membranes: Porin, with threefold symmetry, creates three separate pores, and the potassium channel creates a single pore straight down the axis of fourfold symmetry. The complement C1 molecule at the bottom uses sixfold symmetry to link six identical binding arms, so that it can form a tight multihanded grip on its targets. The four enzymes in the center column show dihedral symmetry, linking 4, 6, 8, and 12 subunits together. In each, the subunits communicate with one another to regulate the action of the protein. In the right column, three examples of cubic symmetry are shown. Tetrahedral symmetry is rare and is used here to link 12 enzyme subunits together. Octahedral symmetry, as in ferritin, and icosahedral symmetry, as in virus capsids, are commonly used to create hollow containers.

economy. To construct a symmetrical complex, molecular assemblers only require enough information to construct one subunit, instead of instructions for the entire assembly. Symmetrical complexes are also economical in terms of error correction. Faulty subunits may be discarded, so that a single error does not ruin an entire assembly, just one of the parts. This is important in natural systems because the error rate of protein synthesis (described in Chapter 2) limits the size of most protein chains to several hundred amino acids. To make larger structures, multiple chains must be combined.

Apart from these synthetic advantages, symmetrical assemblies also have many functional advantages. Assemblies with several subunits can take advantage of *cooperativity,* where the function of one subunit is modified by the state of neighboring subunits. As described in Chapter 5, intersubunit communication is widely used in natural proteins. Symmetry is also used to create a molecule with several identical binding sites, enhancing the strength of binding to its target. The two binding sites in Y-shaped antibodies are a familiar example of this. Finally, symmetry provides a rich palette for creating bionanomachines with useful morphologies, such as calipers, rings, and cups, and molecular infrastructure, such as filaments, membranes, and bulk biomaterials.

In symmetrical complexes, each subunit is identical with its neighbors in structure and environment. Each subunit interacts with its neighbors in an identical way. As described in the section below, biomolecules interact through large complementary patches on their surfaces, which exactly orient one molecule relative to its neighbor. Symmetrical complexes are created by designing a molecule with two or more interacting surfaces that are complementary to one another and designing an appropriate relative orientation of these interfaces (Figures 4-15 and 4-16). With proper design, closed structures with two or more subunits may be created. With a slight modification in the design, unbounded polymeric structures in one, two, or three dimensions are formed.

To create an assembly of defined size and composition, point group symmetries are often used. In point group symmetries, one or more rotational axes pass through a single point, forming a closed, finite assembly. Point group symmetry falls into three general classes of increasing complexity—cyclic, dihedral, and cubic—and is used in the construction of most soluble biomolecules.

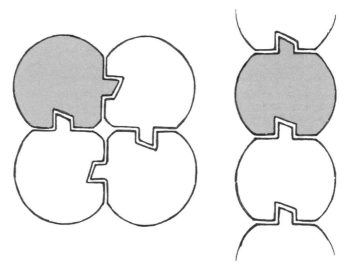

Figure 4-15   By placing the interacting surfaces at different locations, assemblies with different symmetries may be created. On the left, the two interacting surfaces have been placed 90° apart. This subunit associates with three others to form a closed tetramer. On the right, the two interacting surfaces have been placed on opposite sides. This molecule will assemble into a filament.

Cyclic groups contain a single axis of rotational symmetry, forming a ring of symmetrically arranged subunits. Two-fold symmetry is common in natural biomolecules, forming a variety of molecular calipers, scissors, and other functional shapes. The higher cyclic groups are less common. Trimers are created from subunits that have interfaces oriented by 120° relative to one another, tetramers are formed by interfaces related by 90°, and so on with appropriate integral fractions of 360° to form higher-order rings. Bionanomachines with the higher cyclic groups are used in specialized functions where directionality or sidedness is needed, such as interaction with membranes or rotational motion. Assemblies with cyclic symmetry are also useful for functions that require formation of a hollow tube or chamber.

Dihedral groups are characterized by a central axis of twofold or higher rotational symmetry perpendicular to another axis of twofold symmetry. Dihedral complexes have multiple surfaces of interaction, each different. The simplest is D2 symmetry, with four subunits related by twofold axes.

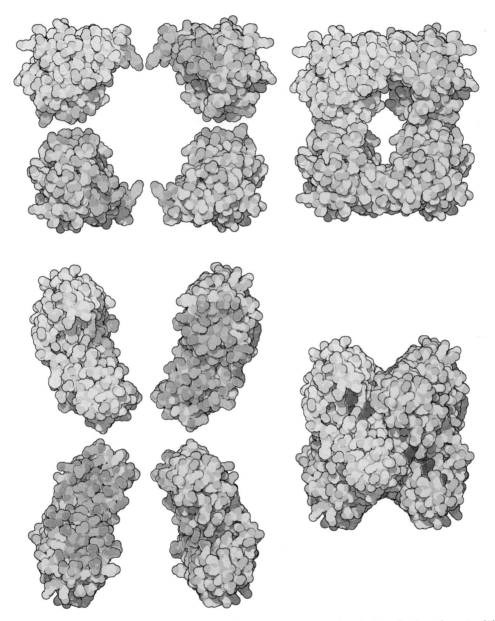

Figure 4-16   Identical symmetry may be used for different functions by placing the interfaces in different locations. These two enzymes are both tetramers with dihedral D2 symmetry. β-Tryptase, at the top, has two small interfaces. When these subunits associate, the tetramer forms a ring designed to protect the active sites, which face inward. On the other hand, the four subunits of phosphofructokinase, shown at the bottom, intimately interlock. As substrates (shown here in bright red) bind, the subunits of phosphofructokinase communicate with one another through these extensive surfaces of interaction, together shifting shape to regulate the activity of the entire complex.

Three separate interfaces must be designed in these complexes, although in many observed cases one of the three possible interfaces actually does not make any contact, forming a ring structure instead of a tight tetrahedron. Higher dihedral symmetries are conveniently thought of as two rings of subunits stacked back-to-back together. Dihedral symmetries, and tetramers with D2 symmetry in particular, are common in natural biomolecules. The different types of interface between subunits on dihedral complexes provide a rich infrastructure from which to build structure and achieve control and communication between subunits. Dihedral symmetries are used widely to construct enzymes that modify their action through communication between subunits.

Cubic groups contain an axis of rotational symmetry with a nonperpendicular threefold axis. Three arrangements are possible: tetrahedral, with two- and threefold axes; octahedral, with four- and threefold axes; and icosahedral, with five- and threefold axes. Cubic groups, because they are composed of so many subunits, often form hollow shells. The design of interfaces for cubic symmetry is exacting, and subunits may be thought of as specially shaped tiles on a spherical surface. In some cases, the subunits are perfectly geometric in shape, with remarkably sharp angles. Cubic groups, with their precise arrangement of symmetry axes, are used in specialized roles to create containers for storage and transport. Icosahedral symmetries are uniquely suited for creating large, hollow shells, such as those used by simple spherical viruses to carry their genetic material.

Translational symmetries in one, two, or three dimensions are used to form extended structures. Most often, translational symmetries are combined with rotational symmetries in biomolecules. Translational symmetries are unbounded, unlike point group symmetries, and may form structures of indefinite size. In biological systems, the growth of these structures is carefully controlled, typically through a set of regulatory biomolecules that cap growing ends or severing proteins that break structures that are too large or have become obsolete.

Line symmetries include a translation in one dimension. Adding a rotational symmetry around the translation axis yields a helix (Figure 4-17). Perpendicular twofold axes may also be incorporated, forming a double helix or higher-order intertwined helices. Before any structures of proteins were

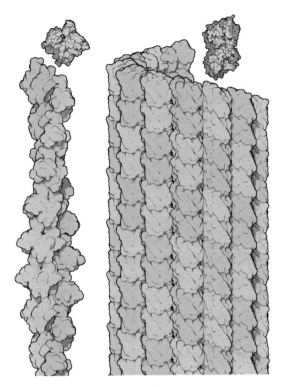

Figure 4-17 Helices are the most common bionanostructures incorporating translational symmetries. A variety of different structural forms may be created by placing the interfaces in different relative orientations. For instance, the subunits of actin, on the left, stack on top of one another, forming a long, thin filament. Tubulin, however, has interfaces that interact around the entire perimeter of the subunit. They stack like bricks in a circular chimney to form hollow microtubules, as shown on the right.

known, Linus Pauling proposed that protein subunits with two complementary surfaces of interaction would assemble into a hollow helical fibril. Since then, many of these structures have been found, including microtubules used for cellular infrastructure, flagella used for bacterial propulsion, and the tubular coat of tobacco mosaic virus. Helical symmetries are also used to build narrower structures, with no central hollow, by orienting binding surfaces such that a small number of subunits form each helical repeat. The actin filaments bracing cells are examples of this.

Plane symmetries are formed when translational symmetries are applied in two dimensions, perhaps with the addition of rotational symmetries. The arrangement of bricks within a brick wall is an example of a plane symmetry. Plane symmetries are relatively rare in biomolecules but may be increasingly important as bionanotechnology strives to create large structures. Natural examples include the plane arrays of S-layer proteins found on the surfaces of some bacteria.

Space group symmetries, with translational symmetries in three dimensions, are rare in natural biomolecular systems. Examples include three-dimensional lattices of collagen in connective tissues and crystalline fibers of sickle cell hemoglobin. Small crystalline arrays are also used to store hormones before release and to store enzymes such as catalase in compact form inside specialized cellular compartments.

Translational symmetries pose a special challenge. Helical assemblies and two-dimensional and three-dimensional lattices have no intrinsic limits on size, and thus can grow without bound. In biological systems, and in applications of bionanotechnology, this can spell disaster. A variety of approaches may be used to specify the size of unbounded polymers. The most obvious approach is to use a scaffolding or ruler molecule of defined size to measure out the proper length (and breadth or width) of the polymer. This approach is used in the construction of muscle sarcomeres: Nebulin measures the actin filaments, and titin measures the myosin filaments. These large protein chains adopt an extended conformation that extends the length of the polymer. Similarly, a three-dimensional scaffolding is used in bacteriophage capsids to specify an elongated shape instead of perfect icosahedral symmetry.

A time-limited approach may also be taken. You can incorporate a time-activated switch that controls whether the assembly is being made or destroyed at any given time. For instance, microtubules assemble and disassemble dynamically, timed by a chemical clock based on the cleavage of a bound nucleotide molecule. Microtubules assemble slowly and then rapidly disassemble when the amount of cleaved nucleotide reaches a critical level. This ensures that a relatively consistent range of lengths is found at any given time. Alternatively, you can simply create the proper number of subunits by timing the duration of synthesis or delivery. Assembly is stopped when the subunits are all used. This approach is useful when a single large structure is being constructed. For example, this method has been proposed as the mechanism for controlling the length of bacterial flagella.

Quasisymmetry Is Used to Build Assemblies Too Large for Perfect Symmetry

Point group symmetries allow the construction of complexes about 10 times larger than an average-sized protein. Of course, much larger assemblies may be constructed if several different protein subunits are combined, but at the cost of specifying the construction of all these different pieces. Viruses are caught between these two extremes: Some viruses need to build very large capsids, but they do not carry enough genetic information to specify many different types of subunits. Instead, they employ *quasisymmetry* to built capsids that are larger than possible with perfect symmetry (Figure 4-18). The basic concept is simple, but the details of design remain a challenge.

Quasisymmetrical complexes are built by placing two or more identical molecules at each symmetrical position. For instance, perfect icosahedral viruses are built with 60 identical subunits but quasisymmetrical viruses are built of 120, 240, and larger multiples of subunits. Quasisymmetry requires that subunits adopt slightly different conformations in the different nonsymmetrical positions. In the original conception of quasisymmetry, it was thought that smooth deformations of subunits would be sufficient to fit subunits into the different nonsymmetrical neighborhoods, much like the slight differences in lengths of struts used by Buckminster Fuller to build large geodesic domes. But the atomic structures of viruses reveal that structural switches are far more common, with subunits adopting two or more stable structures with different orientations of interacting surfaces. One can imagine a hierarchical assembly process whereby several subunits assemble, each with different conformations, to form a protomeric complex, which then assembles with perfect symmetry. Although this model may be more simplistic than what is observed in real assembly processes, it provides a ready model for design.

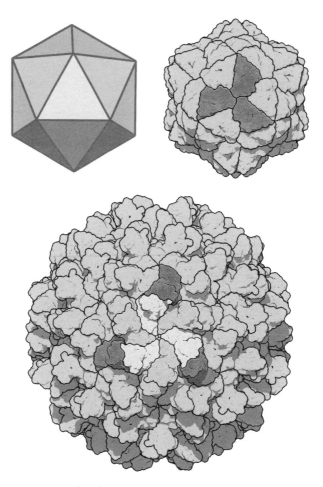

Figure 4-18   Quasisymmetry is used to build structures larger than possible with perfect symmetry. The virus on the top is built with perfect symmetry using 60 subunits. Note how the three subunits in red comprise one of the triangular faces of the icosahedron. The larger virus on the bottom is built with 120 subunits. The subunits are all chemically identical, but the details of how they interact with their neighbors are slightly different. Note that in this virus, nine subunits form each triangular face, and there are three slightly different conformations, in different shades of red and pink.

Crowded Conditions Promote Self-Assembly

Natural bionanomachines are designed for optimal function under crowded conditions. Bionanomachines occupy 20–30% of the volume inside cells. Crowding has profound effects on the assembly and function of bionanomachines. In crowded environments, each molecule is packed next to many others, so the space around the molecule is continually blocked by the presence of neighboring molecules. As you might expect, physical blockage inhibits diffusion, slowing it to a tenth or a hundredth of the normal rate. And, as expected, the diffusion of larger molecules is slowed even more than the diffusion of smaller molecules.

Perhaps not as obvious, crowding also increases the association of molecules into larger complexes. Experimental and theoretical studies have shown that crowded conditions favor compact forms of molecules. Take, for instance, a dimeric protein held together by a weak interaction. In dilute solution, the two molecules spend most of the time apart, associating only transiently. However, if a high concentration of some background molecule is added, the dimeric form will be preferred, even though the background molecule does not interact directly with the dimeric species. The equilibrium is tipped toward the dimeric species by a change in the entropy. Volume-occupied conditions inhibit the diffusion of molecules. So, once the two monomers find one another, they will be shepherded toward one another again and again, strengthening the dimeric form. Thus weak interactions, which may be impossible to observe in dilute conditions, may become important in crowded, cellular contexts. This increase in association is primarily restricted to large molecules.

This association effect can have interesting consequences. For instance, it can change the products formed by DNA ligase, the enzyme that connects the ends of free DNA strands. If DNA ligase is added to a dilute solution of DNA fragments, it will connect them end to end, forming a long DNA helix. If, however, a high concentration of background molecules is added to the mixture, cyclic forms are made instead of end-to-end forms. Under crowded conditions, the DNA fragments have trouble finding other fragments, so DNA ligase tends to connect the two ends of each fragment.

Notice that these two effects—slowing diffusion and increasing association—act in opposite ways when we are looking at the rate of enzymatic re-

---

Lessons from Nature

- Self-assembled structures are modular, require specific geometry of interaction between modules, and, in complex environments, require a unique interaction of subunits.
- Symmetrical modules are often used to reduce the information needed for construction, to provide error control, and to create assemblies with symmetrical functional shapes.
- Point group symmetry may be used to design assemblies of defined size.
- Quasisymmetry may be used to create assemblies that are larger than those possible with exact symmetries.
- Crowded environments slow diffusion but enhance assembly of large molecules.

---

actions. Crowding will slow down the diffusion of substrates to enzymes, but once they are there crowding will tend to favor binding of the substrate to the enzyme and thus will favor reaction. Thus the actual rate of a reaction may be affected in complex ways by the presence of a high concentration of background molecules. For bionanotechnology, it may become necessary to add a background molecule to reproduce the proper environment for a biomolecule-inspired process. Many enzymes are found to perform better when a crowding molecule, such as polyethylene glycol or a neutral protein like serum albumin, is added to the reaction mixture. This ensures that the enzyme adopts the proper state and improves its interactions with any partners present in the solution.

SELF-ORGANIZATION

Self-assembly is perfect for building nanomachinery of precise size and shape. In some applications, however, a less concrete building material is needed. Self-organization is a perfect method for creating structures that are flexible, resilient, and self-repairing. Self-organized systems lack the control that is available with self-assembly, but that very lack of defined structure is what is needed in some applications. In natural systems, self-organization is used primarily to create lipid membranes. In current bionanotechnology, a number of self-organized forms of lipids and lipidlike molecules are being explored to create novel infrastructures and to create delivery vehicles for nanomedicine.

Like self-assembled systems, self-organized systems are modular. They differ, however, in the lack of specific surfaces of interaction. The modules in self-assembled complexes interact with defined geometry, so the final structure is composed of a specific number of modules precisely arranged into a defined assembly. Self-organized systems, on the other hand, have large nonspecific surfaces of interaction, allowing a wide range of similar interactions between neighboring modules. Self-organized assemblies of a given module may adopt a variety of different flexible shapes.

Lipids Self-Organize into Bilayers

Lipids are small molecules composed of a water-soluble chemical group connected to one or more carbon-rich tails. A few variations on this common design, described in Chapter 2, are used in cells, but many other variations on this basic chemical architecture are possible for use when designing a new application. Because of their long hydrocarbon tails, lipids are highly insoluble in water. Self-organization is driven, as in protein folding, by the need to shelter these carbon-rich tails from water (Figure 4-19). Each lipid has a distinctive *critical concentration*. When placed in water at concentrations greater than the critical concentration, lipids associate to shield the hydrophobic segments from water. The critical concentration is very low and is lower for lipids with longer carbon chains. For instance, a typical lipid with 16 carbon atoms in each chain will have a critical concentration in the picomolar range.

The shapes of the individual lipid molecules determine the form of the self-organized aggregate. Wedge-shaped lipids tend to favor the formation of spherical micelles, filling out the sphere like slices of a pie. Cylindrical molecules, such as the abundant phospholipids used in cells, form extended

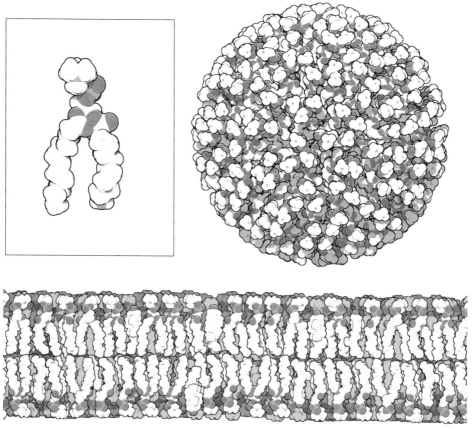

Figure 4-19   Lipids, such as the phospholipid shown at upper left, spontaneously self-organize into micelles, shown at upper right, and bilayers, shown in cross section at bottom. The self-organization is driven by the need to shelter their carbon-rich tails, shown in white, from water.

lipid bilayers. The bilayer is composed of two sheets of lipids, oriented such that the hydrophilic head groups are exposed on the two surfaces of the sheets and the carbon-rich tails extend inward toward the center.

Lipid Bilayers are Fluid

Because lipid bilayers are composed of many nonbonded molecules, they are dynamic structures. Although lipids adopt a variety of crystalline phases under different conditions, under typical biological conditions, the bilayer is fluid and the component lipids are constantly flowing relative to one another. Lateral motion is fast, but flipping of lipids from one face of the membrane to the other occurs only infrequently. The fluidity of lipid bilayers is useful because it allows spontaneous healing of damage. Fluidity also allows rapid communication between molecules within the membrane, as they diffuse and interact with neighboring molecules.

The fluidity of lipid bilayers is dependent on the structure of the component lipids and on the temperature. Bilayers composed of a single type of lipid, and in particular, lipids built with straight saturated chains, tend to crystallize into a frozen state. Lipid chains with double bonds (unsaturated bonds) and mixtures of different lipids both inhibit crystallization and favor fluidity. Cholesterol, used widely in higher cells, is a special case. Cholesterol, when added at significant (5–20%) concentrations, interacts with portions of the hydrocarbon chain closest to the head group, reducing their mobility. This stiffens phospholipid membranes and also decreases their permeability. But, at the same time, it helps prevent the crystallization of the phospholipids.

Biological membranes are highly flexible, allowing complex shape transformations. The remarkable flexibility of red blood cells is an excellent example: They are normally disk-shaped, but they fold to flow through capillaries that are half the diameter of the disk. The act of budding, where small vesicles pinch off from a larger vesicle, and fusion, where several vesicles combine to form one seamless vesicle, are processes that also require significant flexibility and fluidity of the membrane. On the other hand, membranes are resistant to forces in the plane of the membrane. A change in surface area of roughly 1 percent will cause rupture of the membrane, which is strongly resisted by the hydrophobic forces holding the membrane together.

Proteins May Be Designed to Self-Organize with Lipid Bilayers

Of course, a perfectly sealed membrane is useful for only the most limited applications. Methods are needed to communicate and transport materials across the membrane. In natural systems, these functions are performed largely by specialized proteins that interact with the membrane. Membrane proteins associate with lipid bilayers in a variety of useful manners (Figure 4-20). In the simplest case, a lipidlike group may be attached to a protein surface. This lipid will insert into the bilayer, tethering the protein to the surface of the bilayer. Many other proteins are designed with hydrophobic segments that span the membrane or insert on one side of the membrane. Some contain a single stretch of approximately 20 amino acids of predominately hydrophobic character. These form an α-helix, shielding the hydrophilic peptide groups inside and displaying the hydrophobic sidechains outside. This structure will pass through the membrane perpendicular to the plane, exposing the two ends of the protein on opposite sides of the membrane. Many other proteins are built with a hydrophobic girdle that orients the protein within the membrane, like a bather surrounded by an inner tube.

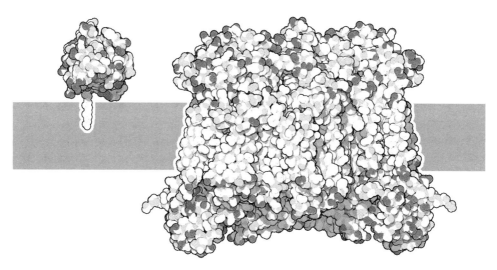

Figure 4-20 Proteins interact with membranes in different ways. On the left, the ras protein has a short hydrocarbon chain attached to the protein chain. This tail inserts into the membrane and tethers the protein to the surface. Cytochrome oxidase, shown on the right, is a large globular complex of proteins that spans a lipid membrane, shown schematically as a gray stripe. The protein is designed with a ring of carbon-rich amino acids that surrounds the protein, orienting it in the membrane. Note that the portions that extend into the water above and below the membrane are studded with charged atoms, shown in bright red here.

Lessons from Nature

- Self-organizing systems are composed of modules that interact through nonspecific patches on their surfaces.
- Lipids self-organize into a variety of aggregates based on their structure, including micelles and bilayers.
- Lipid bilayers are fluid, flexible, and self-repairing.
- Proteins may be designed to self-organize with lipid bilayers, either by addition of a lipid or by tailoring a lipidlike surface on the protein.

Because the membrane is fluid, these proteins rapidly diffuse in two dimensions along the membrane. Many biological processes, described in Chapter 5, rely on diffusion and interaction of proteins within membranes.

MOLECULAR RECOGNITION

Most macroscale machinery is composed of a number of interacting parts. An automobile includes many separate parts rigidly fixed together with screws, rivets, welds, adhesive, snap fittings, and other fasteners. Articulated parts move by way of precisely machined axles, pistons, and joints. The freedom of motion between these articulated parts is modified according to function, to reduce friction in wheel axles or to increase friction in brake pads.

Most nanomachines will also include a number of interacting parts, so methods of *molecular recognition* are needed to control the interaction between parts. Unfortunately, the many principles used in macroscale engineering are not readily applicable at the nanoscale. In particular, the atomicity of nanoscale objects poses a major challenge. Nanoscale objects typically interact through direct contact of only a few dozen atoms. At the nanoscale, we do not have the range of material types and the ability to shape components to arbitrarily high tolerances that we enjoy with macroscale machinery. We are limited to a few atom types and their interactions through dispersion, hydrogen bonds, and electrostatic forces. We are further limited by the geometry of the bonds linking the atoms in each molecule. We cannot design a nanoscale component of any arbitrary shape. Each must be made of an integral number of atoms, connected according to the simple rules of covalent bonding.

Crane Principles for Molecular Recognition

Before the first atomic structures of biological molecules were determined, the physicist H. R. Crane postulated that two design concepts would be required for macromolecular recognition in self-assembling systems (Figure 4-21). First, "for a high degree of specificity the contact or combining spots on the two particles must be multiple and weak." This may not seem obvious: We might think that it is better to use one very strong interaction to hold two parts together. Using one or a few strong interactions will provide stability. However, it will not provide specificity. The same arrangement of a few strong combining sites might be found on many other molecules, increasing the risk of improper pairings. Instead, an array of many weak interactions is better. Then, all of the interactions are necessary to add up to the proper binding strength.

Second, "one particle must have a geometrical arrangement which is complementary to the arrangement on the other." The shape of the interacting surface must form a tight fit, bringing the "multiple, weak interactions" into the proper alignment. In biological molecules, this complementarity includes matching shapes, with knobs on one molecule fitting into holes on the other, and matching chemical groups forming the proper hydrogen-bonding pairs and pairs of charges across the interface between molecules. This close complementarity is essential for the specificity of interactions. A stable interface could be created with many gaps, but this interface could form stable interactions with other generic interfaces. Thus knobs are important for fitting into the proper holes, but also for bumping into improper surfaces and disallowing binding to them. Overlap of a single methyl group can provide enough unfavorable energy to disallow binding of an entire protein-protein interface, whereas a single missing hydrogen bond will not destabilize the interaction enough to preclude binding. A proper design

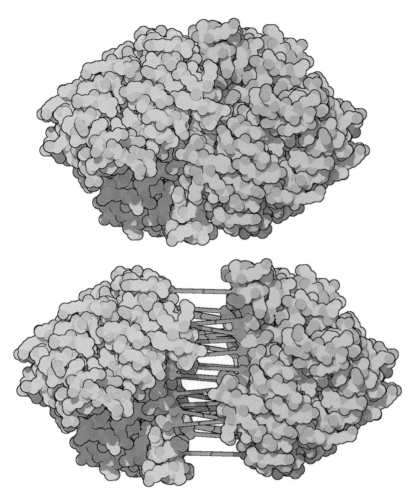

Figure 4-21   The Crane principles are shown in this structure of enolase, at the top. The active enzyme is composed of two identical subunits that associate into a tight dimer. On the bottom, the dimer is separated and cylinders are drawn to show all of the hydrogen bonds that are formed between the subunits. Note the complementarity of the hydrogen bonds and the matching of the overall shape of the interface.

will have a unique arrangement of these many weakly interacting atoms for each pair of interacting proteins, ensuring that each is different enough to exclude binding to competing molecules.

These two design principles have been observed in hundreds of examples of natural bionanomachines. Biomolecules interact through extended surfaces, with many weak sites of interaction arranged in a perfectly complementary shape. Nearly all of the interactions involved in molecular recognition are noncovalent. Direct covalent bonds connecting two subunits are rare. Covalent bonds are far too restrictive in the biological environment and are only used when extremely sturdy structures are needed. Instead, a combination of hydrogen bonds, electrostatic charge interactions, and burial of hydrophobic regions is used.

The interfaces in natural protein assemblies show a wide range of solutions to this problem. Roughly one-third of protein-protein interfaces show a classic form, stabilized by forces similar to those that stabilize protein folding. At the center of the interface is a carbon-rich patch, often with a convoluted shape. When two subunits associate into the dimer, this patch provides hydrophobic stabilization and energetic impetus for dimerization. The shape is designed to fit only with the proper partner. Around this central hydrophobic patch, there is a ring of hydrogen-bonding and electrostatic interactions from protein to protein, stitching up the perimeter of the interface surface. Often, individual water molecules also bridge hydrogen bonding groups between the two proteins.

The remaining proteins show a more free-form interface, with a patchwork of carbon-rich hydrophobic regions, hydrogen bonds, and bridging water molecules. These are distributed like a crazy quilt across the entire interface, coming into perfect register when the two proteins associate in the proper conformation.

Similar principles are used in the recognition of small molecules by proteins. The small size of these molecules, however, provides fewer atoms for interaction but allows a more intimate interaction. Typically, ligand-binding sites are formed in a deep pocket or even buried completely, accessible only through a large opening-and-closing motion of the protein. These pockets are designed to recognize a given molecule, using a array of hydrogen-bonding groups and carbon-rich hydrophobic pockets that exactly complement the chemical character and shape of the ligand.

Adenine recognition provides an example (Figure 4-22). Bionanomachines that use adenine nucleotides must recognize the adenine ring but at the same time exclude guanine nucleotides, which are similar in chemical characteristics and also common in the cellular environment. This is done by a combination of perfectly placed hydrogen bonds and strategic physical

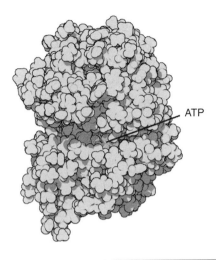

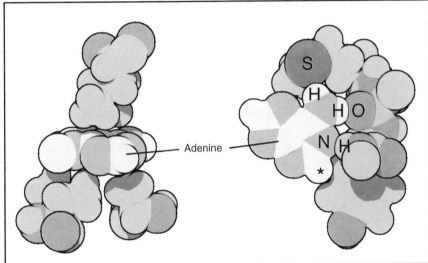

Figure 4-22 Adenine is recognized with a form-fitting active site and a special contact that excludes the binding of guanine. The signaling protein cAMP-dependent protein kinase is shown here. Looking at the amino acids that interact with the adenine ring, several methods are used to recognize the ring. As shown on the right, several carbon-rich amino acids pack on the top and the bottom of the ring, forming a slot that is perfectly shaped for the flat adenine ring. As shown on the right, several amino acids form hydrogen bonds with atoms in the ring, between hydrogen and oxygen, nitrogen, and even sulfur. Note how this protein also packs a tyrosine amino acid tightly against the hydrogen atom marked with a star. Guanine has a bulky amino group attached at this position, which would be blocked by the tyrosine if it tried to bind in this active site.

blocking. Adenine-binding sites are typically deep, slot-shaped pockets, tailored to fit the disk-shaped molecule. Complementary hydrogen bonding groups are arrayed around the edges of this groove, serving to register the base within the slot. Hydrogen bonding, however, does not provide a way to exclude guanine from the site. A different approach is taken. Guanine has a large amino group in a place that is empty in adenine. Within the binding site, one or more atoms are placed in tight contact with this site, so that the larger amino of guanine cannot fit. So, even if the hydrogen bonding sites that are common to both adenine and guanine are used, the specific blocking group will provide specificity for adenine.

Specific blocking may be used to disallow binding of competing molecules that are larger than the target molecule, but how can nanomachinery reduce binding of smaller competing molecules? Or molecules with slightly

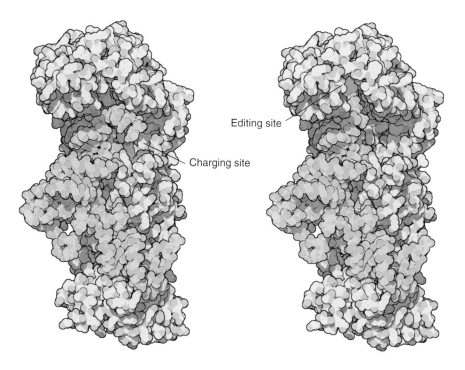

Figure 4-23   Proofreading is used to improve the accuracy of leucyl-tRNA synthetase, the enzyme that connects leucine to its proper tRNA during protein synthesis. The enzyme, shown in gray, contains two active sites. The charging site adds leucine to the tip of the tRNA, shown in pink. However, it also occasionally adds valine instead of leucine. The second active site is an editing site that only fits valine, clipping it off when it is improperly added.

mispaired hydrogen bonds? The energy of interaction will be smaller, given that the number of contacts is reduced, but this may not be enough to ensure exclusive binding of the desired molecule. This is a significant problem in bionanomachinery, which is solved in some cases by a separate proof-reading step that is performed after two molecules interact.

The enzymes that connect amino acids to transfer RNA are faced with this problem. The enzyme leucyl-tRNA synthetase has the job of attaching a leucine amino acid to the proper tRNA (Figure 4-23). The active site has a small pocket that fits the 4-carbon leucine side chain perfectly, positioning it properly for attachment. However, the 3-carbon side chain in valine also fits fairly well in this site and is attached about 1 time in 200 instead of the proper leucine amino acid. This problem is solved by incorporating a second active site that cleaves the amino acid from the tRNA. It is built with a smaller pocket, which fits the valine but strongly excludes the proper leucine. Thus, through this "double-sieve" mechanism, improper connections are destroyed, increasing fidelity by 10 times.

Proofreading is also used in other places to improve the accuracy of natural protein synthesis. DNA polymerase is one example. It copies the genetic information in DNA, separating the strands and creating a new copy to pair with each. As each new base is added to the growing strand, it must be chosen correctly to complement the template strand, always pairing adenine with thymine and always pairing cytosine with guanine. This job must be performed with high precision, despite the fact that the proper pairings of bases are only slightly more favorable than some of the improper pairings. DNA polymerase adds a separate proofreading step to the normal synthetic reaction to improve its accuracy. DNA polymerase contains two active sites. One site matches each new base with its partner on the template strand and connects it to the growing strand. The other site then pulls on this new base, and, if the interaction is weak, it cleaves the base off. This extra step removes bases that are improperly paired and ultimately increases the accuracy a thousandfold.

Atomicity Limits the Tolerance of Combining Sites

Macroscopic machinery may be constructed with high precision. Component parts may be fashioned to fit together perfectly, often with tolerances

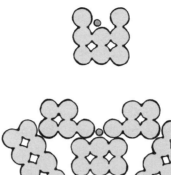

Figure 4-24   The precision of bionanomachines is limited by the atomicity of molecules. As an example, look at a simple binding site composed of a square lattice of atoms, which is designed to recognize a smaller atom. A hole cut into a square lattice has a precision of about an atomic radius, forming a poor recognition site for the small atom. Bionanomachines, however, use the approach shown in the bottom structure. They achieve higher tolerances in specific locations by using a large overhead of surrounding infrastructure to position a few key sites.

of micrometers, or one part in a million. This allows immense flexibility in the design of macroscopic machines. Construction may be performed with screws, dovetails, and a variety of other fasteners. Moving parts include axles, hinges, gears, etc. Information is stored by changing the color or magnetic properties of small patches of material or physically with small dents or bumps.

The molecular level does not allow nearly this precision. Nanomachinery must be composed of individual atoms, each roughly spherical and limited by the geometry of covalent and nonbonded interactions (Figure 4-24). Surfaces of interaction are rough; molecules cannot be designed to make a seamless fit between surfaces. For instance, consider the construction of an arbitrary shape with carbon atoms on a diamondoid lattice. The atomicity of the lattice would limit the precision to about an angstrom, but no better. This lack of precision in the fitting of molecular parts limits the precision of any recognition or sensing event.

Natural bionanomachines, however, achieve much higher precision in key portions of a structure. The folding of protein chains can place a dozen

---

Lessons from Nature

- Recognition surfaces are complementary in shape and chemical properties and contain many weak, specific interactions.
- Physical blocking may be used to exclude binding of larger competing molecules, and proofreading may be used to improve recognition when faced with smaller competing molecules.
- Atomicity limits the range of properties and tolerance of interacting molecules.
- Natural bionanomachines combine a stable, global structure of arbitrary precision and a few local regions specified to high precision.

---

atoms in precise positions. Hemoglobin provides an example (see Figure 5-15). Oxygen binding induces a 0.5-Å motion of an iron atom out of the plane of a heme group. A histidine residue senses this small motion of the iron atom and transmits it to the rest of the protein complex, magnifying the motion, through use of a structure switch, into a large conformation change.

Proteins optimized by evolution do not demand precision in every interaction. Instead, they are optimized such that a few critical contacts are placed in a precise location. The bulk of the protein is used as an infrastructure that creates a precise interaction site. This is one of the great challenges in protein design: predicting the proper balance of the many forces that stabilize protein structure in order to create this local precision. The selection method of evolution is a perfect method for discovering bionanomachines with this combination of properties: a stable, global structure of arbitrary precision and a few local regions specified to high precision.

FLEXIBILITY

Much of our macroscopic technology is based on rigid components machined from metal, plastic, or other materials that resist deformation. When flexible motion is needed, an articulated joint, a spring, or another locally flexible component is designed and used to connect the rigid components.

Much of molecular nanotechnology, because it is based so closely on macroscopic engineering, is taking this same approach, designing rigid components that interact through a few mobile bonds or joints. The natural bionanomachines selected by evolution, on the other hand, are intrinsically flexible. Evolution starts at the opposite extreme, building structures that are highly flexible and then selecting molecules with the necessary level of rigidity needed for function.

Flexibility is a key design feature of biomolecules. Many of the processes performed by biomolecules, and enzymatic catalysis in particular, occur in conformations that are significantly distorted from the equilibrium structure of lowest energy. Bionanomachines might be likened to rubber tires or plastic pop beads as opposed to precision-machined axles or machined gears. Flexibility is a great asset in biomolecules, allowing subtlety in action and the resilience that is a hallmark of life. It poses great challenges, however, as we attempt to design new bionanomachines that harness this flexibility.

Biomolecules Show Flexibility at All Levels

Biomolecules show flexibility across the entire range of scale, from the atomic level all the way to the level of assemblies. At the smallest scale, small vibrations along bond lengths and bond angles occur continually under the force of thermal energy. These fluctuations can be critical in the chemical reactions catalyzed by enzymes. A temporarily stretched bond may be weakened, speeding a chemical transition, or the angle between two bonds may be strained to promote transition to a different geometry.

The motion of side chains, loops, or entire domains at the surface of biomolecules is used extensively in molecular recognition. This has been dubbed *induced fit* because the surface of one molecule conforms to the topology of its interacting partner. Induced fit allows binding sites to surround a binding partner (Figure 4-25). Local changes in side chains can improve the binding of a ligand, and motion of entire domains can enfold a ligand, shielding it from water and burying it in an entirely hydrophobic environment.

Stable folded proteins undergo *breathing* motions, where whole α-helices or β-sheets shift, breaking and forming hydrogen bonds and other in-

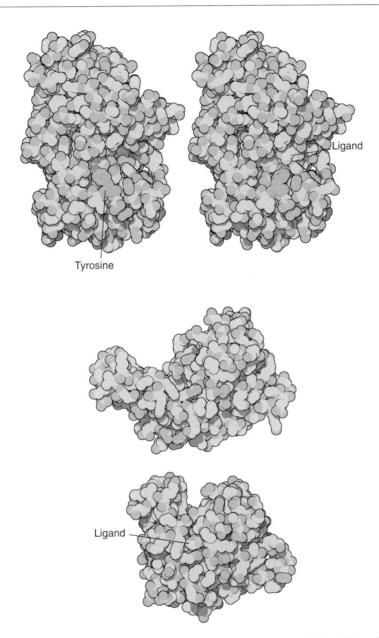

Figure 4-25   Enzymes often use induced fit to improve the interaction with their ligands. In thermolysin, shown at the top, tyrosine amino acids create flexible doors that open and close as substrates bind. Adenylate kinase, shown at the bottom, closes around its substrates, shielding them from water. The reaction involves the transfer of phosphate from ATP to AMP. If water were present in the active site, the phosphate could easily be transferred to a water molecule, essentially cleaving ATP. The active site machinery of adenylate kinase only becomes active when the different domains of the protein are brought together in the closed form, when water is not present.

teractions transiently. These motions are essential, for example, for the diffusion of substrates into buried active sites. Looking at the crystallographic structure of hemoglobin, for example, there appears to be no path for oxygen to reach the binding site near the iron. Oxygen reaches the binding site by pushing through the small transient channels that open through breathing motions.

Many proteins are composed of several relatively rigid domains connected by flexible linkers. These proteins bend and flex, opening and closing like clams. Flexible linkers play important roles in multidomain proteins, particularly those involved with large-scale regulation and binding to large targets (Figure 4-26). Numerous examples may be found in the immune system, where the target is a foreign cell surface, and in DNA information control, where linkers are used to connect specific binding elements to signaling elements that interact with adjacent molecules. These linkers

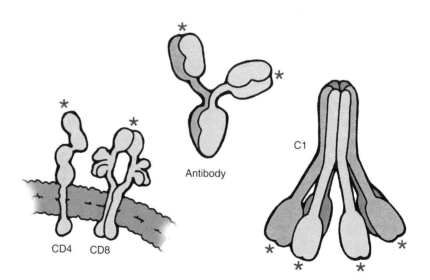

Figure 4-26 Flexibility is essential for biomachines that interact with large targets. Many molecules from the immune system show a characteristic structure composed of several binding domains, shown here with red stars, connected by flexible tethers. The two CD molecules extend from the surface of immune system cells and must be flexible to seek out their targets on invading cells. Antibodies and the C1 protein have several binding sites tethered together, allowing them to bind to several adjacent sites on their target.

have special amino acid sequences, often rich in glycine and proline, to reduce the amount of folded, stable structure that they can adopt.

Finally, multichain assemblies may shift between several specific conformations, each with different functional properties. In enzymes, this is termed "allosteric" motion and is used extensively for regulation. Allostery is described in more detail in Chapter 5. Large protein switches are also used in viruses to transition between tight, stable structures that are used outside the cell during infection and more open structures once the virus has entered the cell. These motions often involve structural switches between domains connected by flexible linkers, similar to the switches used in quasisymmetrical associations.

Biomolecules may incorporate specific levels of rigidity to improve the entropy of self-assembly or function. As described above, heat-stable proteins are created by increasing the rigidity of surface loops, inhibiting the unfolding of the protein structure. An essential example of rigidity is the planar peptide linkage connecting amino acids in proteins. This rigidity, combined with a few key contacts between neighboring amino acids, limits the flexibility of protein chains and favors a few secondary structures, such as α-helices and β-strands. This is essential for proper folding and assembly of proteins, limiting the range of possible conformations of these complex molecules. Nucleic acid chains are far more flexible, but they are ruled by a different constraint: the flat planar bases. The combination of stacking of the bases and the specific pairing of bases limits the range of structures.

Lessons from Nature

- Flexibility at all levels is used to enhance the function of bionanomachines. This includes harnessing of thermal motion for chemical catalysis, use of induced fit for recognition, design of different conformational states for use in regulation, and incorporation of selective flexibility to link several separate functionalities.

Flexibility Poses Great Challenges for the Design of Bionanomachines

These many-layered motions add levels of complexity, providing many options for the optimization of function. However, this flexibility adds an extra level of difficulty when designing bionanomachines. We need to be able to predict the breathing and flexing motions of the molecule as well as the stable equilibrium structure. Evolutionary design, on the other hand, is perfectly suited to design and optimization of this type of action. Small modifications are made step-by-step, arriving at the optimal structure rather than predesigning it. Biomolecular flexibility will provide one of the greatest challenges, and potential benefits, of bionanotechnology.

FUNCTIONAL PRINCIPLES OF BIONANOTECHNOLOGY

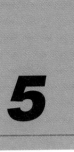

5

Atoms at a small scale behave like nothing *on a large scale.*

—Richard Feynman

The natural tools available in living cells—ribosomes, enzymes, DNA—give us the ability, today, to build working nanomachines. Now we can start to think about what we can do. But, as many visionaries of nanotechnology are finding, design of function at the nanoscale is a formidable challenge. Fortunately, we have abundant examples to explore as we begin to design our own nanomachinery.

Nanoscale function is where biology excels. We can find natural bio-nanomachines performing nearly every imaginable function. Looking in cells, we can find working assemblers, sensors, motors, factories, rigid and elastic materials, adhesives, and the list goes on. And quite remarkably, all this is created with a remarkably limited set of building blocks.

It is easy to be daunted by the complexity of natural bionanomachines. Because they were developed through evolution, they have an unfamiliar organic architecture, not the parsimonious lines of our familiar machinery. But many of the basic functions of bionanomachines reduce to the proper positioning of a few key atoms. Whether these are arranged on a protein architecture or a diamondoid lattice, the function will remain the same.

Bionanotechnology: Lessons from Nature. By David S. Goodsell
ISBN 0-471-41719-X   Copyright © 2004 John Wiley & Sons, Inc.

INFORMATION-DRIVEN NANOASSEMBLY

Information-driven synthesis is essential for any practical technology. To make a technology successful, we must be able to assemble our available raw materials into many final products, based on instructions held in a blueprint. Otherwise, we must create a new assembler for each new product, which would pose too great an investment of resources to allow practical development of new machines and new methods. Once the investment is made in producing an information-driven assembler, anything may be constructed simply by giving it the appropriate instructions.

This principle is used throughout macroscopic engineering. With the guidance of a blueprint, thousands of different buildings are constructed from a standard set of bricks and boards. Computers and radios are constructed from standard electrical components based on a schematic. Molecular nanotechnology hopes to build an assembler that directly presses together the smallest components, individual atoms, based on a plan. Life on Earth, and much of bionanotechnology, relies on an assembler that builds proteins based on information stored in DNA.

This assembler, the ribosome, produces proteins of any length and composition based on a set of instructions. Information-driven protein synthesis is a complex process requiring the concerted action of hundreds of nanomachines. Cells contain molecular tools for storage, editing, duplication, and repair of the molecular blueprints and molecular tools for using this information to build proteins. The system is robust and may be isolated from cells to perform this custom assembly under our control. Alternatively, cells may be engineered to use their existing synthetic machinery to construct proteins for our specific needs. Either way, this assembly mechanism gives us the ability to construct large quantities of custom-designed nanomachinery.

Nucleic Acids Carry Genetic Information

In the natural process of protein synthesis, the blueprint for building proteins is held in DNA strands (Figure 5-1). The information is encoded in the sequence of the four bases in DNA. The code is very simple (now that we can look in retrospect) and universal on Earth. Information is transferred from one nucleic acid strand to another through the specific interaction of

```
caagcaggtc tgttccaagg gcctttgcgt caggtgggct cagggttcca gggtggctgg
accccaggcc ccagctctgc agcaggggagg acgtggctgg gctcgtgaag catgtggggg
tgagcccagg ggccccaagg cagggcacct ggccttcagc ctgcctcagc cctgcctgtc
tcccagatca ctgtccttct gccATGGCCC TGTGGATGCG CCTCCTGCCC CTGCTGGCGC
TGCTGGCCCT CTGGGGACCT GACCCAGCCG CAGCCTTTGT GAACCAACAC CTGTGCGGCT
CACACCTGGT GGAAGCTCTC TACCTAGTGT GCGGGGAACG AGGCTTCTTC TACACACCCA
AGACCCGCCG GGAGGCAGAG GACCTGCAGG gtgagccaac cgcccattgc tgcccctggc
cgcccccagc caccccctgc tcctggcgct cccacccagc atgggcagaa ggggcagga
ggctgccacc cagcaggggg tcaggtgcac tttttaaaa agaagttctc ttggtcacgt
cctaaaagtg accagctccc tgtggcccag tcagaatctc agcctgagga cggtgttggc
ttcggcagcc ccgagataca tcagagggtg ggcacgctcc tccctccact cgcccctcaa
acaaatgccc cgcagcccat ttctccaccc tcatttgatg accgcagatt caagtgtttt
gttaagtaaa gtcctgggtg acctggggtc acagggtgcc ccacgctgcc tgcctctggg
cgaacacccc atcacgcccg gaggagggcg tggctgcctg cctgagtggg ccagacccct
gtcgccagcc tcacggcagc tccatagtca ggagatgggg aagatgctgg ggacaggccc
tgggagaag tactgggatc acctgttcag gctcccactg tgacgctgcc ccggggcggg
ggaaggaggt gggacatgtg ggcgttgggg cctgtaggtc cacacccagt gtgggtgacc
ctccctctaa cctgggtcca gcccggctgg agatgggtgg gagtgcgacc tagggctggc
gggcaggcgg gcactgtgtc tccctgactg tgtcctcctg tgtccctctg cctcgcgct
gttccggaac ctgctctgcg cggcacgtcc tggcagTGGG GCAGGTGGAG CTGGGCGGGG
GCCCTGGTGC AGGCAGCCTG CAGCCCTTGG CCCTGGAGGG GTCCCTGCAG AAGCGTGGCA
TTGTGGAACA ATGCTGTACC AGCATCTGCT CCCTCTACCA GCTGGAGAAC TACTGCAACT
AGacgcagcc tgcaggcagc cccacacccg ccgcctcctg caccgagaga gatggaataa
agcccttgaa ccagccctgc tgtgccgtct gtgtgtcttg ggggccctgg gccaagcccc
acttcccggc actgttgtga gccctccca gctctctcca cgctctctgg gtgcccacag
gtgccaacgc cggccaggcc cagcatgcag tggctctccc caaagcggcc atgcctgttg
gctgcctgct gcccccaccc tgtggctcag ggtccagtat gggagcttcg ggggtctctg
```

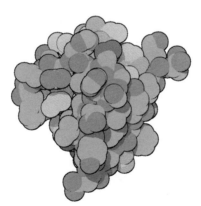

Figure 5-1 The segment of the human genome with the gene for insulin is shown at the top. The region coding for the insulin protein is shown in red and is coded in two parts separated by a noncoding segment. The protein is constructed in a longer form, encoded by the DNA sequences shown in capital letters, and then cut to size after synthesis. The first segment, in black capital letters, is a temporary targeting signal that directs insulin to be excreted from cells. The two red segments are the code for the mature protein. The central segment in black is necessary for the proper self-assembly of the protein and is discarded after the protein matures. The final "TAG" is the stop signal at the end of the coding region. The mature protein, with two chains, is shown at the bottom. The chain corresponding to the first segment in the DNA sequence is shown in pink, and the second chain is gray. The bright red atoms are sulfur atoms that form a bridge linking the two chains.

the bases: adenine with thymine or uracil and guanine with cytosine. This transfer of information is direct: As described in Chapter 2, nucleic acids are replicated by aligning nucleotides directly along a single strand and then connecting them together. In natural systems, transfer of information from one nucleic acid into a new strand is performed by *polymerase* enzymes. These enzymes move step-wise along a strand of nucleic acid and pair nucleotides to each base in turn, creating a complementary strand. *RNA polymerases*, like the one shown in Figure 2-9, create a strand of RNA with DNA as the template, and *DNA polymerases* create a strand of DNA with DNA as the template. Viruses contain other enzymes that transfer information in nonstandard ways: *RNA-dependent RNA polymerases* create RNA based on an RNA template and *reverse transcriptases* create DNA based on an RNA template.

Several properties are needed to create an effective polymerase. The first is accuracy. Many polymerases are extremely accurate in copying. For instance, the DNA polymerase in bacterial cells makes only about one mistake in a trillion nucleotides added, while building new strands at rates of up to 1000 nucleotides per second. In a bacterial cell, this is about one error in a thousand generations. The specificity of the chemical interactions between bases, however, is only sufficient to provide an error rate of one mistake in 10,000 to 100,000 bases copied. As described in Chapter 4, the additional accuracy is provided by proofreading schemes.

Polymerase action is also vastly improved if the enzyme is processive. When a polymerase binds to a template strand, it should copy large stretches before the strand is released. In many natural polymerases, processivity is increased by incorporation of a "clamp" that surrounds the DNA, allowing a sliding motion along its length but not release. In the DNA polymerase used for bacterial replication, this increases the processivity by 500,000 times over polymerases that do not have clamps. Note that the ribosome is also a clamp that closes around the RNA message, allowing it to read the entire length of an RNA strand before releasing it.

A full toolbox of natural enzymes has been characterized that modify and interact with the information held in nucleic acids. Most are available commercially and are routinely used to perform these functions in the laboratory. These include:

(1) *Polymerases,* as described above, create new nucleic acid strands based on the information in other strands.

(2) *Nucleases* cut nucleic acid strands. These include nonspecific nucleases that cut in many places and specific nucleases such as restriction enzymes that cut only at specific base sequences.

(3) *Ligases* connect nucleic acid strands.

(4) *Repressors, transcription factors, enhancer proteins,* and other regulatory proteins bind to nucleic acids and regulate the use of the information. They may act by physically blocking use of the strand, by providing an entry point for binding of polymerases, or by modifying the physical structure of the DNA.

(5) *Base-excision nucleases* remove bases from nucleic acid strands. These are often specific for modified bases or mismatched bases and are used to correct errors.

(6) *Topoisomerases* solve the topological problems encountered by long strands of DNA. This includes relaxing overwound DNA, forcibly overwinding DNA, and allowing one portion of the strand to pass through other portions to untangle DNA.

(7) *Recombinases* extract a portion from one strand of DNA and swap it with a similar portion in a different strand.

(8) *Spliceosomes* edit a piece of RNA, removing a portion in the center and reconnecting the two flanking segments.

(9) *Nucleosomes* and other proteins package nucleic acids for storage.

As described in Chapter 3, the availability of these molecular tools has spawned an entire discipline of recombinant DNA technology that allows researchers to modify DNA and use it to create custom proteins and engineered organisms.

Of course, when looking at natural systems, everything is not quite as simple as we might expect. The information stored in DNA is not used directly by ribosomes to build proteins. Instead, an intermediary molecule is used. For use, the information is copied, or *transcribed,* into a strand of RNA. This RNA strand is then used directly by the ribosome to direct the construction of proteins. From a design standpoint, the need for the intermediary RNA step is not obvious, but all living cells go through this step (however, some viruses skip DNA entirely, carrying their genetic information in RNA and simply reproducing it). Perhaps this is another example of an evolutionary legacy. Many researchers believe that life was developed with RNA and the more stable DNA molecule was added only later, when the genetic information was refined enough for archiving. However, the extra step does provide abundant opportunity for regulation, an opportunity used in many ways in cells and available for use in various aspects of bionanotechnology.

Ribosomes Construct Proteins

Ribosomes are nanoassemblers that construct proteins according to the information in RNA (Figure 5-2). The process is complex, because the genetic code is an informational code and not a chemical code. The transcription of

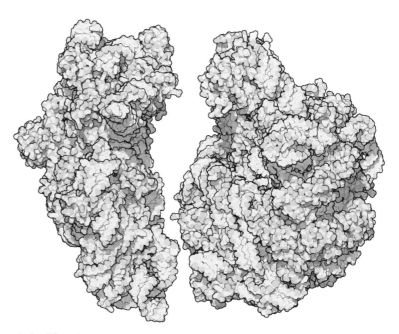

Figure 5-2 The ribosome is an information-driven molecular assembler. It is composed of two subunits, each constructed both of RNA, shown in pink, and of proteins, shown in gray. Together, they close around the messenger RNA that carries the instructions for building a protein.

DNA to RNA uses a chemical code: Each DNA base chemically matches an RNA base. All that RNA polymerase needs to do is display the DNA strand and connect the RNA bases when they bind properly to the strand. The translation of RNA to protein, on the other hand, is not based on chemical interaction between the RNA and the protein amino acids. Instead, it is a translation of one chemical language into another. And, as with all translations, a dictionary is needed for matching words in one language with their proper counterparts in another. In all cells, the dictionary is composed of *transfer RNA*, a collection of translator molecules made of RNA. These L-shaped adapters have a segment at one end that recognizes the information in RNA and a segment at the other end that is chemically connected to the proper amino acid.

In natural systems, the genetic code is a sequential, contiguous triplet code. A *codon* of 3 nucleotides is used to specify 1 of the 20 amino acids, with a few codes reserved for "Start" and "Stop" directions. Codons are aligned sequentially along the RNA strand with no intervening bases. The code is degenerate, as there are more codon combinations than amino acids. As described in Chapter 6, some researchers are modifying this code for specific nanotechnology applications.

The ribosome is a complex machine with many moving parts. It wraps around a strand of messenger RNA and steps one codon (three bases) at a time down its length (Figure 5-3). At each codon, it aligns the proper transfer RNA to the codon. Then, at the other end of the transfer RNA, it attaches the amino acid to a growing protein chain. The process is assisted by dozens of initiation proteins to get it started, elongation proteins that add the necessary chemical energy and push the ribosome forward at each step, and termination proteins that specify when to stop.

The atomic structures of the ribosome provide a detailed look at a working molecular assembler, designed to create custom, atomically precise molecules according to defined instructions (Figure 5-4). It is a daunting structure, all the more unfamiliar because it is composed primarily of RNA instead of protein. However, it appears to have all of the functions one might expect. The small subunit has a clamping mechanism that entraps and positions the RNA strand. The RNA threads through a hole between the two subunits in the complex, promoting a processive synthesis from be-

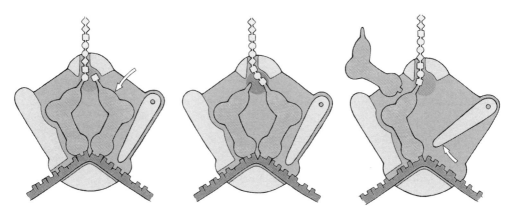

Figure 5-3 The ribosome steps down the RNA strand one codon at a time. The addition of one amino acid to a growing protein strand is shown schematically here. The process begins when a new transfer RNA enters the ribosome and binds to the next codon in line. Then the growing protein chain is transferred to the amino acid held by this transfer RNA. Finally, the transfer RNA is pushed one step forward, ejecting the previous, now-empty transfer RNA.

ginning to end. The synthetic step is performed by a typical active site, although surprisingly using chemical groups from RNA for the catalysis. Addition of each new amino acid is performed by forcible displacement of spent transfer RNA by elongation factor proteins, using ATP cleavage for power.

Ribosomes are accurate in their synthesis, but not as accurate as information transfer between nucleic acids. In bacteria, improper amino acids are added with a frequency of about 1 in 2000 amino acids. About one in four typical proteins (of about 500 amino acids) will have an error. However, most missense errors are fairly harmless, as most single-site mutations in a protein will only slightly compromise its function. However, processivity errors, where protein synthesis is terminated prematurely, are more significant. In bacteria, the frequency of premature termination is about once in 3000 codons, so about one in seven typical proteins is released before it is fully synthesized. Larger proteins are even more difficult to construct in full.

Information Is Stored in Very Compact Form

The flow of information from DNA to RNA to protein demonstrates that information may be densely stored at the nanoscale. The direct transfer of in-

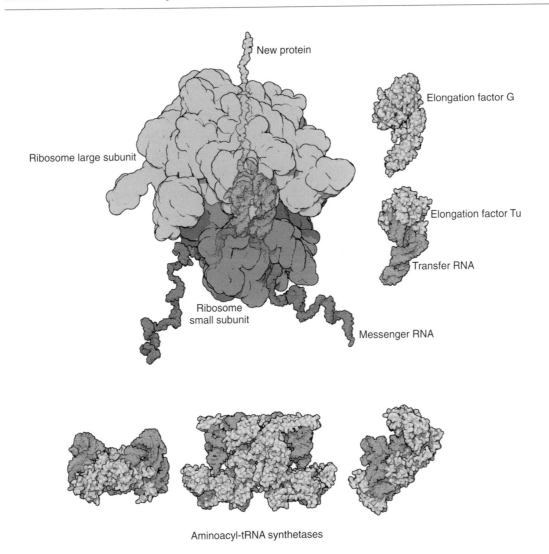

Aminoacyl-tRNA synthetases

Figure 5-4 Structures involved in protein synthesis. The ribosome is shown at the top. The small subunit positions the strand, and the large subunit pairs transfer RNA molecules along the message strand and perform the synthesis reaction. Elongation factor Tu delivers new transfer RNA molecules to the ribosome, and elongation factor G performs the job of the lever shown in Figure 5-3: It forcibly moves transfer RNA molecules one step forward after the synthesis reaction is performed. The enzymes shown at the bottom connect the proper amino acids to each transfer RNA—only 3 of the full set of 20 are shown here.

Lessons from Nature

- Information-driven synthesis allows construction of a diverse collection of products with a single construction mechanism.
- The ribosome is a working molecular assembler that constructs a linear polymer by using a linear information-storage medium. It provides an existence proof for information-driven atomic nanoassembly.
- Processivity of synthesis can be improved by use of clamps that surround information-carrying strands.
- Information may be densely encoded in specific arrangements of atoms. A working two-bit code is used in DNA, requiring an overhead of about 15 atoms per bit.

---

formation by chemical interaction allows very compact storage and read-out. Each position in a DNA strand carries two bits of information, because each nucleotide can adopt one of four states, corresponding to the four bases. An overhead of about 30 atoms is needed to position each base in such a way that the information may be stored, copied, and read. This is a remarkable information density: 15 atoms per bit.

Jonathan Cox has discussed the challenges of using DNA as a long-term archival storage medium. When suitably protected, DNA may be stable for millions of years. For instance, DNA held in bacterial spores can be revived after long periods of time. However, retrieval of information held in DNA is cumbersome and a million times slower than modern computer information storage. But it has a great advantage over very long time spans. Current information technology becomes obsolete in a manner of decades, replaced by faster, denser technologies. DNA, however, will probably retain its value as an information storage medium for millions of years. Because it is tied so closely to living systems and because evolutionary legacy will most likely ensure that DNA will be the storage material used in living systems, the processes of DNA information storage and retrieval will be available in nearly identical form for millennia.

ENERGETICS

Many desirable nanoscale processes do not occur spontaneously. In these cases, we must add energy to force the process to occur in the way that we want. Fortunately, there are many other highly energetic processes, such as the breakage of chemical bonds, the capture of light, or the reuniting of separated charges, that we can harness for powering processes that are more sluggish. Looking at natural bionanomachinery, we can find examples that use all three of these sources of energy—chemical energy, light energy, and electrical energy. These sources of energy are used in two main ways: to drive difficult chemical reactions and to power directed motion.

Living cells, however, do not harness energy like we do in the macroscopic world. We typically create a large quantity of heat and then use this to power motion. For example, think of an automobile engine. The explosive burning of gasoline, a favorable chemical process, powers the moving of the engine. Cells, on the other hand, do not use reactions that release a great deal of heat to perform their mechanical or chemical work, because thermal energy is rapidly dissipated throughout nanoscale systems before it can be captured for use. (However, heat may be produced in the body on demand, either physically through friction in the rapid motion of muscles or chemically by increased rate of heat-producing reactions such as the breakdown of fat molecules.) Instead, energy is metered out in small steps, so that it can be controlled and efficiently captured.

Natural bionanomachines transfer energy by linking two processes together. For instance, two chemical reactions may be linked together, using a very favorable reaction to boost a less favorable one. Take, for example, the reaction performed by the enzyme pyruvate kinase. It performs two separate reactions: It removes a phosphate from phosphoenolpyruvate and adds a phosphate to ADP. When performed separately, the first reaction is highly favorable. The phosphoenolpyruvate molecule is unstable, but the two pieces that form when it is broken—pyruvate and free phosphate—are each highly stable. The second reaction, on the other hand, is very difficult. It is difficult to connect another phosphate to the end of ADP because of the strong charges on the phosphates that are brought into close contact. When linked together, however, the combined reactions are slightly favorable, and the entire process will occur spontaneously. Similarly, chemical reactions may be linked to electrical processes, or the capture of light can be used to power chemical reactions, or other combinations may be employed. The key is to transfer the energy in nanoscale pieces.

Chemical Energy Is Transferred by Carrier Molecules

One of the most common approaches for powering chemical reactions in natural systems is to link an unfavorable reaction to a second, highly favorable reaction. You might imagine that this linkage could be approached in two different ways. We could approach each new task as a new challenge, trying to discover a new combination of complementary reactions each time. Or we could develop a common fuel molecule that we could link to any unfavorable reaction that we choose. In cells, the latter approach is by far the most common approach. Of course, this simply adds an intermediary to the process—we must create a mechanism for creating these fuel molecules and then develop methods to link them to our ultimate reactions. The modularity of the system, however, is a significant advantage.

Looking at the fuel molecules used in cells, we find that they are all built with a similar design. They are composed of an energy-transferring group attached to a convenient nanoscale handle. The energy-transferring group relies on chemical instability. Many familiar molecules are unstable and are useful for capturing energy. Acetylene has an unstable carbon-carbon triple bond at its center, which breaks when it combines with oxygen to produce a very hot flame. TNT and nitroglycerin have nitrogen and oxygen atoms poised next to carbon and hydrogen atoms—one wrong move and the molecules rearrange, exploding into a more stable cloud of nitrogen gas, carbon dioxide, and water. These compounds are difficult to construct but easy to destroy. The molecules used in cells, however, are not this extreme. They are unstable because of charges that are brought into unfavorable contact, or they might contain atoms that are frozen into unfavorable bonding states. It is difficult to build these unfavorable linkages, and when they are allowed to break they can be used to drive other processes.

The handles attached to these unstable reservoirs of energy are designed to be recognized by the nanomachines that use the fuel molecule. Molecules like acetylene and TNT do not have handles, so they are only useful for creation of energy in bulk. Biomolecular fuel molecules contain handles built of moderately sized organic compounds, allowing the fuel to be manipulated one molecule at a time. These handles typically have a large number of oxygen and nitrogen atoms, allowing nanomachines to use specific hydrogen bonds to recognize them.

ATP (adenosine triphosphate) is the most common biological fuel molecule. Several methods are used to construct ATP with energy from the breakdown of food or the capture of light. Cleavage of ATP is then used to power most unfavorable biomolecular processes. ATP is an unstable molecule with a close connection between phosphate groups. Each phosphate carries a strong negative charge, so it is difficult to bring them together and very favorable to let them separate. The chemical energy trapped in the unstable ATP bond is spent in many ways. It is used to ensure that key chemical reactions are performed when needed, even if they are not normally favorable. ATP is used for a variety of mechanical processes as well, where the shape or location of a molecule must be forcibly changed.

The adenine ring provides the handle for recognition (Figure 5-5). As described in Chapter 4, enzymes recognize ATP by using a combination of shape and chemical complementarity. Typically, the adenine ring binds in a deep pocket that recognizes the flat, planar shape. Hydrogen bonding groups are arrayed around the perimeter of this pocket, positioned to form hydrogen bonds with the amino groups on the ring. This positions the ATP

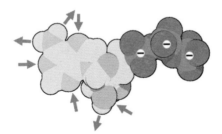

Figure 5-5 ATP contains an adenine ring at one end, shown on the left here, that serves as a convenient molecular handle for recognition. It has several sites for hydrogen bonding, shown with arrows. Three phosphates are linked directly together at the other end, on the right here. Each of these phosphates carries a negative charge. Breakage of the phosphate-phosphate bonds, allowing the negatively charged groups to separate, is a favorable energetic process and is used to power other unfavorable processes.

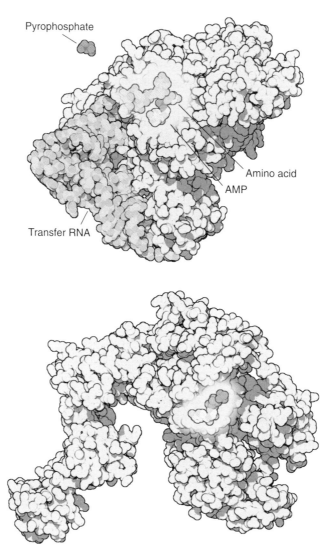

Figure 5-6 ATP may be used to power many different nanoscale processes. The enzyme aspartyl-tRNA synthetase, shown at the top, is using ATP to power the addition of an amino acid to a transfer RNA. This structure shows an intermediate stage in the process. The enzyme begins by bringing together a transfer RNA, ATP, and the amino acid. Then, as shown here, it connects the amino acid to the ATP, releasing two of the phosphate groups in the form of pyrophosphate. Finally, it transfers the amino acid to the transfer RNA. The breakage of ATP provides an energetic boost to this normally sluggish reaction. ATP is also used to drive the power stroke of the muscle protein myosin, as shown at the bottom. ATP binds in the middle of myosin and controls actin binding through the cleft on the right and the forcible motion of the long lever arm on the left.

in the proper position for transfer of its energy when the phosphate-phosphate group is broken (Figure 5-6).

Light Is Captured with Specialized Small Molecules

Nearly all life on Earth is ultimately powered by light from the Sun. The light-capturing event is performed by a class of proteins termed *photosynthetic reaction centers*. These proteins capture a photon of light and use it to create a high-energy electron, which is then used for power. The reaction center contains a series of cofactors—chlorophyll, phylloquinones, and iron-sulfur clusters—that do the work. A special pair of chlorophyll molecules absorbs the photon, exciting an electron into a higher-energy state (Figure 5-7). Normally, this excited electron would lose energy through heat or would emit a photon of slightly lower energy as fluorescence. But the reaction center is designed to circumvent these normal avenues. Instead, the excited electron is quickly transferred away from the chlorophyll along the chain of cofactors.

Ultimately, this high-energy electron is placed on a carrier molecule to be transferred to the site of usage. The missing electron from the initial chlorophyll is replaced by a low-energy electron from a second source. The result is the transfer of electrons from a low-energy source to a high-energy carrier. In most photosynthetic organisms, the electron is obtained from water, which is oxidized to oxygen gas in the process. The electron is promoted to higher energy and then placed on metalloprotein carriers for delivery to other processes.

Photosynthetic organisms also contain effective molecules that harvest light and transfer it to reaction centers (Figure 5-8). These proteins are packed with chlorophyll and carotenoid molecules that absorb light of many wavelengths. The energy is then transferred from molecule to molecule by resonance energy transfer, until it reaches the special pair of chlorophylls in the reaction center, where the excited electron is quickly shuttled away.

Energy from light is also harnessed to do physical work. For instance, the protein bacteriorhodopsin transports protons across a membrane by using power provided by the absorption of light, and the light-sensing protein

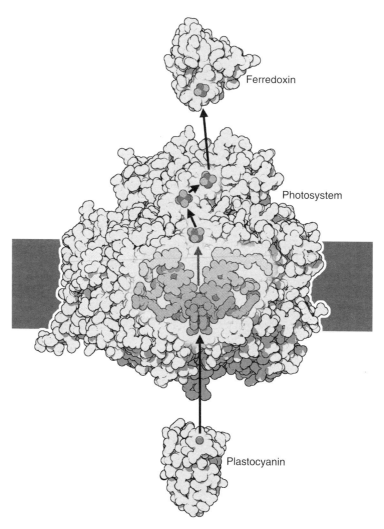

Figure 5-7 Photosynthetic reaction centers use a pair of chlorophyll molecules to absorb light and activate electrons. The photosystem from a cyanobacterium is shown here, with special chlorophyll molecules in dark pink at the center. A series of electron-carrying prosthetic groups then carry the activated electrons through several chlorophylls and a phylloquinone (shown with a red arrow) and through three iron-sulfur clusters. Ultimately, the electron is placed on a soluble carrier protein like the ferredoxin at the top. Plastocyanin, shown at the bottom, then replaces the missing electron with an electron of lower energy.

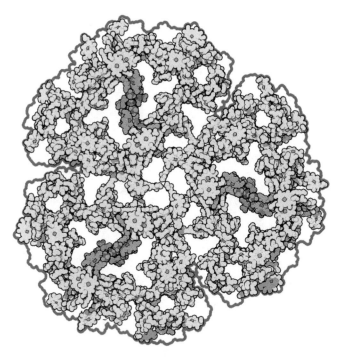

Figure 5-8 Photosynthetic reaction centers often use large arrays of light-absorbing molecules to act as an antenna for gathering light. The same photosystem shown in Figure 5-7 is shown here from the top. The photosystem is composed of three identical subunits, each with its own electron transfer chain in the center, shown in bright pink. Surrounding each are dozens of chlorophyll molecules that absorb light and transfer the energy to the chain at the center.

opsin changes shape when it absorbs light. These are described in more detail below.

Protein Pathways Transfer Single Electrons

Electronics play an enormous role in macroscale technology. Electrical conduction is an example of *charge transport*, where electrons are flowing in bulk. The metal atoms in the wire allow electrons to move freely, even in the absence of an external force. If no external potential is applied the electrons diffuse randomly, but if a voltage is applied there is a net motion of electrons. Electrical conduction is a novelty at the biological nanoscale. As

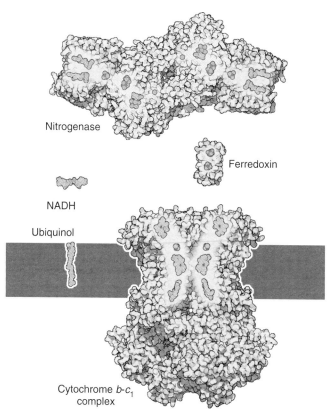

Figure 5-9   A wide variety of electrical components are used to create electrical bio-nanocircuits. These include small organic molecules that transfer electrons from one site to the next. Some are water soluble, such as NADH, and others, like ubiquinol, are insoluble and shuttle electrons inside lipid membranes. Small proteins, such as ferredoxin, are also used to shuttle electrons from one site to another. These small carriers are used to deliver electrons to many large electric-powered bionanoma-chines, such as large membrane-bound proton pumps like the cytochrome b-c_1 complex and enzymes like nitrogenase that perform reduction reactions. These proteins use a variety of prosthetic groups to manage the flow of electrons.

described below, DNA is a conductor of electrons, but there is no evidence that this conduction is used for any function. It comes as no surprise that cells use a more controlled approach to electronics.

Biological systems move electrons one at a time from one carrier to the next in well-defined bionanocircuits. This process is termed *charge transfer*. The transfer of single electrons along complex paths is widespread in bio-

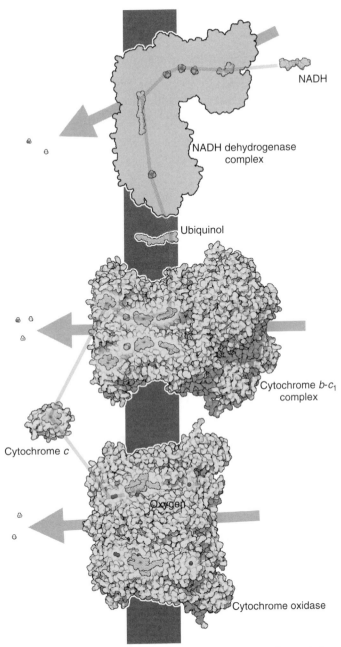

NADH

NADH dehydrogenase
complex

Ubiquinol

Cytochrome b-c_1
complex

Cytochrome c

Oxygen

Cytochrome oxidase

Figure 5-10   The *electron transport chain* is a bionanocircuit that links the flow of electrons to formation of a proton gradient. High-energy electrons are obtained by the oxidation of glucose and carried to the chain by NADH. The electrons then flow through three large membrane-bound proton pumps. They are transferred from one complex to the next by small, mobile carriers: ubiquinol and cytochrome c. As the electrons flow along the chain of prosthetic groups in the three large protein complexes, the energy of the flow is used to power the transfer of protons across the membrane, as shown schematically by the large gray arrows. Ultimately, the electrons are deposited on oxygen molecules, converting them to water.

logical systems, so we have many powerful examples to use as templates for building our own single-electron bionanocircuits (Figures 5-9 and 5-10). In these pathways, individual electrons are transferred between specific carrier molecules. A variety of different prosthetic groups are used to carry the electrons, including iron-sulfur clusters, copper ions, iron ions held in heme groups, and polycyclic organic ring systems. In addition, small, mobile organic molecules, such as NAD and FAD, can be used to deliver electrons between proteins in these pathways.

Each electron carrier is characterized by a reduction potential that quantifies its affinity for electrons. The protein chains surrounding specific prosthetic groups can tune the potential of the group by orders of magnitude, by positioning key amino acids that stabilize or destabilize the binding of electrons. With proper design, the potential of each carrier in a pathway may be chosen to provide an ordered pathway from start to finish. Driven by a spontaneous reduction in free energy, electrons flow in progression from one carrier to the next along the circuit.

Electrons are transferred from carrier to carrier by quantum mechanical tunneling, which is effective for distances of up to about 1.4 nm. If electrons must be transmitted over longer distances, chains of carriers are used, with each step less than about 1.4 nm apart. Transfer rates are very fast at these distances, in the range of 10^{13} to 10^7 per second. A recent survey of electron-transferring proteins in the Protein Data Bank by researchers at the Johnson Research Foundation revealed that electron transfer is not strongly affected by differences in the specific amino acids in the intervening space. In their words: "There has been no necessity for proteins to evolve optimized routes between redox centers. The transfer is remarkably robust, as long as the electron carriers are close enough, so design efforts can focus on tuning of each individual redox center and integration of the chain with its inputs and outputs."

Thus far, natural electron transport seems to be limited to two major uses: for bulk delivery of electrons for use in chemical reduction reactions and as a mechanism for powering other processes, such as the pumping of protons. Amazingly, natural systems have not exploited single-electron transport for computation. Biological computation is performed at the nanoscale by hard-wired genetic and biochemical networks and at the microscale by programmable nerve networks. Single-electron computers, however, are an exciting possibility for bionanotechnology.

Electrical Conduction and Charge Transfer Have Been Observed in DNA

DNA contains many aromatic bases stacked one atop the next. The quantum mechanical π orbitals of these bases overlap, creating a pathway for the flow of electrons. DNA is a potential candidate for the design of nanoelectronic devices, with several advantages. The synthesis of DNA is routine, and customized assemblages, anything from single nanowires to complex networks, may be designed and synthesized. This is a beautiful idea in concept, but the details of electrical conduction and transfer in DNA are still hotly debated. Several processes have been observed.

Charge transfer of single electrons has been extensively studied with planar molecules that interact with the DNA helix. These molecules are activated by light and remove an electron from one of the nearby nucleotides, creating an *electron hole*. Researchers then follow the transfer of this hole along the helix to distant sites. Often this is quantified by appearance of the charge at sequences with multiple guanine residues, which lose electrons easily and tend to capture the electron hole. The charged guanine is sensitive to cleavage by chemical reagents, which are used to quantify the amount of the transfer. Analysis of the kinetics of this transfer has revealed two processes: a superexchange mechanism that falls off strongly with distance and a multistep hopping mechanism that covers longer distances.

Conduction of electrons by DNA has been studied by bridging two electrodes with a short DNA helix or bundle of DNA strands and then measuring the current through the strands when a potential is applied. In one experiment, researchers showed that DNA can support significant currents, applying 100 nA (about 10^{12} electrons per second) through a single DNA molecule 10 nm in length. Experiments from other laboratories, however, have shown different results, showing insulating, semiconductive, or even superconductive properties. The isolation of single molecules and the details of the connection between the electrode and the DNA are major challenges that influence results.

Lessons from Nature

- Unfavorable nanoscale processes may be performed by linking them chemically or physically to energetically favorable processes.
- Chemical energy may be stored and transferred by fuel molecules with chemically unstable covalent bonds.
- Energy from light may be captured by light-absorbing prosthetic groups such as chlorophyll to create high-energy electrons.
- The directional flow of single electrons in bionanocircuits is controlled by choice of electron carriers with ordered reduction potentials.
- Electrons are transferred between carrier groups by quantum mechanical tunneling over distances of up to 1.4 nm.
- Electrochemical gradients created across membrane-enclosed spaces may be used to power chemical and physical processes.

Electrochemical Gradients Are Created Across Membranes

Nanoscale energy may be stored by using a concept similar to batteries and capacitors. The idea is to separate charged objects into two separate compartments, so that one holds more negative charge and one is more positive. In a capacitor, electrons are pumped from one metal plate to the other, building up a negative charge on one side. Then the flow of electrons back can be used to power an electrical machine. In cells, ions are typically used instead of electrons. An enclosed space is created, surrounded by a membrane that is impermeable to ions. Then ions are pumped across the membrane, creating an electrochemical gradient. The flow of ions back across the membrane is then used to perform chemical or mechanical work.

Electrochemical gradients provide power in two ways. First, a simple concentration gradient is created. As ions are pumped across the membrane, the concentration increases on one side. Second, an electrical potential is also created as charges accumulate on one side of the membrane. In both cases, work may be performed as ions are allowed to flow backward,

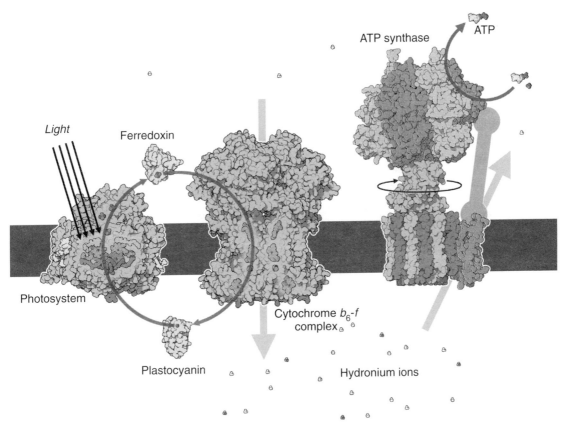

Figure 5-11   The process of cyclic photophosphorylation, which captures light energy to make ATP, combines many of the energetic modalities used in nature. Light is captured by a photosystem and used to create high-energy electrons. These are transferred by ferritin to a large proton pump (cytochrome b_6-f complex) that creates an electrochemical proton gradient powered by the flow of the electrons. The electrons, now at lower energy, are transferred back to the photosystem by plastocyanin to await the next photon. The proton gradient is then used by ATP synthase for the mechanochemical synthesis of ATP. The protons flow back across the membrane, turning the rotary motor portion of ATP synthase in the process. As this motor turns, it forces a change in shape of the enzyme portion of ATP synthase, providing the power to connect ADP and phosphate, forming ATP.

restoring an equilibrium both in concentration and charge distribution across the membrane. These two forces together combine to provide a strong force that is used to power many biological processes (Figure 5-11). The most widespread application is the ubiquitous use of proton gradients. Protons are pumped across membranes, creating a *proton-motive force* that is widely used to turn motors or create ATP.

CHEMICAL TRANSFORMATION

Chemists have been performing specific chemical transformations for centuries, creating structurally pure molecules for use in medicine and industry. This process typically proceeds through a sequence of specific chemical modifications, such as adding a new group, changing a specific bond, or making other small changes. At each step, a new reaction must be designed, often with significant side products competing with the desired reaction. The chemist must purify the desired product at each step and search for reactions that maximize the proper products. For a organic molecule with dozens of atoms, like many of the molecules used as drugs, this may involve dozens of steps and the final yield may be quite low.

Biological machinery excels in one ability above all others: performing specific chemical transformations. Like chemists, cells create specific molecules by an ordered set of small chemical transformations. But, unlike the chemist, they build specific bionanomachines—*enzymes*—that perform each step efficiently and accurately (Figure 5-12). Typically, each enzyme is optimized to perform a single reaction, speeding the chemical process by a tril-

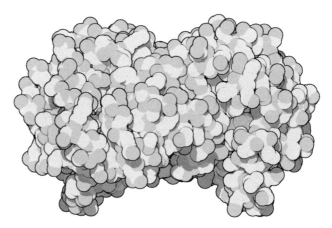

Figure 5-12A Triose phosphate isomerase is an example of a perfect enzyme. It is a diffusion-limited enzyme, performing its reaction at rates faster than the rate at which substrate molecules can diffuse to it. It uses all the tricks employed by enzymes to perform their reactions. The enzyme is a dimer of two identical subunits, each with a separate active site that surrounds the substrate molecule, shown in pink. Note how large the enzyme is relative to the substrate. All this infrastructure is needed to ensure that a handful of key active site amino acids are perfectly arranged.

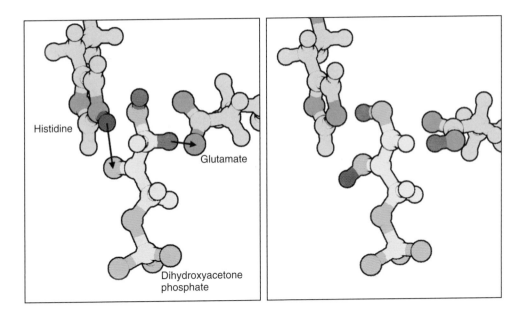

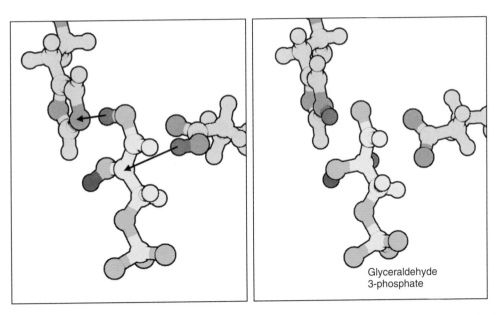

Figure 5-12B Triose phosphate isomerase performs an isomerization reaction, removing two hydrogen atoms (shown in red) from dihydroxyacetone phosphate and replacing them in different positions to form glyceraldehyde 3-phosphate. The reaction is performed in two steps, using two key amino acids. In the first step, shown at the top, a glutamate extracts one hydrogen atom and a histidine adds another (shown in dark red) back to the molecule in a different position. Then, in the second step, the glutamate replaces its hydrogen atom in a different position and the histidine grabs another hydrogen atom from the substrate. Note that the enzyme starts and ends in the same form—with a free glutamate and with hydrogen bound to the histidine (albeit a different hydrogen atom). This leaves the enzyme ready to perform the same reaction on the next substrate molecule that it encounters.

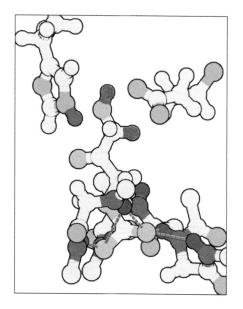
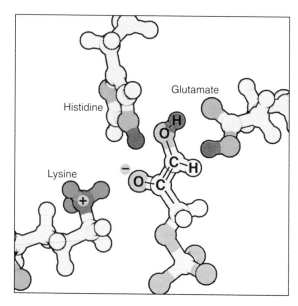

Figure 5-12C The phosphate group of the substrate is surrounded by a collection of amino groups from the protein, shown here in dark gray on the left. They form a specificity pocket that forms hydrogen bonds with the substrate, positioning it correctly relative to the catalytic histidine and glutamate amino acids above. The enzyme speeds the reaction by stabilizing the transition states of the two steps. The transition state of the first reaction is shown here on the right. The glutamate has extracted one hydrogen, but the histidine has not yet donated its hydrogen. This leaves an unfavorable negative charge on one oxygen atom in the substrate. This is stabilized by a lysine amino acid from the enzyme, which carries a positive charge.

lionfold or more over the unassisted rate. The formation of unwanted side products is reduced nearly to zero, so that the desired product, even if it requires dozens of synthetic steps, is created at high yield (Figure 5-13). These machines are also carefully regulated, so that products may be created when and where they are needed. Enzymes are a priceless gift from nature, providing the starting point for all of bionanotechnology.

Enzymes perform chemical transformations by paving the way through the desired reaction, smoothing over any obstructing hills and lowering any roadblocks. A chemical reaction is similar to the process of breaking a pencil. At the beginning the pencil is perfectly solid and static. Then you start to apply pressure and it bends, resisting and straining all the way. Finally, it snaps, and you are left with two pieces, solid and static. A chemical cleavage reaction is similar. If you are breaking a molecule in two, you must pull it forcibly apart. The beginning and final forms—the molecule and the two halves—are perfectly stable, but the intermediate states, as individual bonds are stretched, are highly unstable and energetically unfavorable.

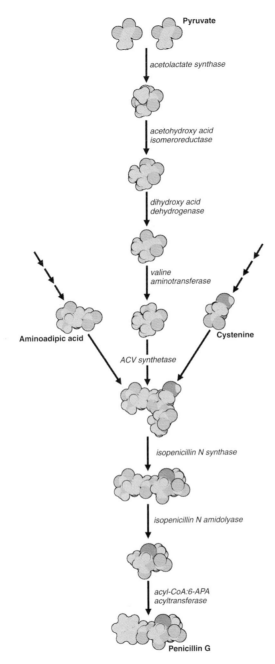

Figure 5-13 Cells and synthetic chemists both build organic molecules in a series of chemical transformations, taking available starting materials and making chemical changes until the desired product is obtained. The refinement added by natural systems is the use of enzymes at each step that provide specificity and efficiency that are not available in the solution processes typically used in organic chemistry. This is the sequence of steps used in bacteria to build penicillin from the common small molecule pyruvate. Two pyruvate molecules are combined and converted into the amino acid valine in four steps. Then it is combined with two other amino acids, each also created by a number of steps from simple precursors, to form the basic skeleton of penicillin. Three additional steps create the active form.

Enzymes reduce the energetic barrier imposed by these intermediate states—termed *transition states*—making them easier to form from the starting material and easier to convert into the proper products. Think again of the pencil. This time, take your fingernail and make some deep dents in the wood along the side. Now, the pencil bends and breaks far more easily. You have catalyzed the pencil-breaking reaction by changing the intermediate states with your fingernail, making the partially bent states easier to achieve. Similarly, enzymes create a molecular environment in which the transition states are stabilized so that they do not present such a barrier to the reaction.

All of the action occurs in the *active site* of an enzyme. There, specific amino acids are placed in strategic locations, perfectly positioned to stabilize the transition state of the molecule undergoing the reaction. Many diverse methods are used, each tailored for a given reaction. At first glance, every active site seems to be different, each developed separately by evolution for its task. But, several general principles are used in most cases: reduction of entropy, chemical stabilization of the transition state, and use of specialized chemical tools.

The three-dimensional structures of hundreds of enzymes have been solved and are available through the Protein Data Bank (http://www.pdb.org). This is a primary resource for bionanotechnology, providing a wealth of working examples of specific chemical catalysts. These provide an excellent starting point for the development of custom enzymes, tailored for nonbiological applications.

Enzymes Reduce the Entropy of a Chemical Reaction

Enzymes speed reactions by having everything at the right place at the right time. Entropy is constantly diluting reactions, reducing the probability that molecules will meet and react in the desired way. In a reaction that requires the joining of two molecules, entropy will ensure that they rarely find one another. In a reaction that requires an exact alignment of two bonds, entropy will ensure that this alignment is just one among many other random alignments. Enzymes are nanoscale jigs that fight entropy by positioning reacting molecules and forcibly aligning reacting bonds in the proper orientation.

Active sites conform closely to the shape of the molecules being transformed. The surface of the active site is complementary to the molecule. It will have carbon-rich patches abutting carbon atoms in the molecule and hydrogen-bonding atoms in perfect registration with hydrogen-bonding atoms on the molecule. Enzymes commonly make contact with most of the molecule, and some enzymes completely enclose their targets, using flaps and doors that close after the molecule is bound.

The active site is typically separated into two functional regions. First, there is a specificity pocket that recognizes the proper substrate and binds tightly to it. Second, there is the catalytic machinery that performs the chemical transformation. A separate specificity pocket, often comprising most of the active site, is needed in most cases because the catalytic machinery is often chemically exotic and must be optimized to perform the chemical reaction instead of providing the optimal binding characteristics.

The typical tolerances for complementarity between enzymes and their substrates are very fine, fitting together to a fraction of a nanometer. These high tolerances make enzymes highly specific in the reactions that they catalyze. Natural enzymes routinely separate molecules that differ by a single atom. They also can be highly stereospecific, separating right-handed and left-handed forms of a molecule or creating only one of many possible forms.

Enzymes Create Environments That Stabilize Transition States

After substrates are locked comfortably into the form-fitting, reduced-entropy active site, enzymes create a chemical environment that promotes the desired chemical transformation. This is the heart of nanotechnology, where specific atoms are removed or added according to demand. Enzymes catalyze reactions by stabilizing the intermediate, transition state of the reaction. This is accomplished in many ways, tailored to the given reaction.

In some cases, chemical groups, provided by surrounding amino acids or prosthetic groups, interact with the substrate, modifying its electronic structure to make portions more reactive. A charged amino acid can polarize a neighboring bond in the substrate, making a target atom more susceptible to attack or making it a stronger attacker. A hydrogen bond may stabilize a form of the substrate that is normally found with low probability but is the right shape for the desired reaction.

In many cases, the transition state may include an unstable charged form of the molecule, with one or more atoms in a less than ideal bonding state. These are often stabilized by placing an amino acid nearby that carries the opposite electronic charge, forming a stable electrostatic interaction with the transition state.

Enzymes can also introduce geometric strain in substrates. In cases in which the geometry of the substrate changes during the reaction, the active site is designed to fit more tightly to the shape of the transition state than to the initial substrates, favoring the transformation from substrate to transition state.

Enzymes Use Chemical Tools to Perform a Reaction

Most enzymes use specific chemical tools to interact directly with the substrate, directly making chemical changes. It is important that these tools end up, after the reaction is performed, in the same state that they begin. This is the definition of a catalyst, which may undergo a chemical change to assist the reaction, but which must be restored at the end so that it is ready to perform subsequent reactions.

Enzymes commonly use reactive amino acids to make specific chemical changes in substrates. For instance, many enzymes use key amino acids to shuffle hydrogen atoms within a molecule. In particular, histidine is often used in this role. At the pH of a typical cell, it is fairly easy to remove one of the hydrogen atoms on histidine amino acids and later replace it. Histidine is used in many reactions that remodel molecules. A histidine in the enzyme will pull a hydrogen atom off the molecule and then replace it in a slightly different position, flipping the handedness of the molecule or moving the location of a double bond.

Other reactions require a more forceful approach. In these enzymes, an amino acid attacks the substrate, forming a covalent bond. Typically the bond is quite unstable, and a subsequent step will break the bond, forming the desired product. Serine proteases are one example, where a serine amino acid attacks a peptide substrate, breaking the chain and forming an unstable bond to one half. Soon after, a water molecule enters and separates the remaining half, restoring the serine to its original form.

Lessons from Nature

- The entropy of chemical reactions may be reduced by assembling the molecules in the proper location and conformation.
- Specific binding sites may be used to define the specificity and stereochemistry of a given reaction.
- Reactions may be catalyzed by creating a specific chemical environment that stabilizes the transition state.
- Specific prosthetic groups may be used to transfer atomic groups to molecules.

In some reactions, enzymes add new atoms to a growing molecule. When you are going to build a new deck at your house, you rarely start from raw materials. You don't chop down trees and mine iron ore; rather, you go to the lumberyard and buy two-by-fours and nails. Similarly, enzymes often build new molecules with prepackaged atoms that are easy to add to a growing product. Many molecules serve as carriers for specific atoms (Figure 5-14). ATP carries a phosphate group, and other molecules are carriers for carbon atoms, sulfur atoms, nitrogen atoms, and hydrogen atoms. These molecules are cleverly designed: All hold their atoms with an unstable bond, so they are easily released to their new position. These coenzymes are often exotic-looking molecules that must be synthesized specially for the job.

Some reactions simply require the shuffling of a few electrons. Metal ions typically play the role of electron carriers, because they can cycle between several stable charged forms. Copper and zinc ions are often held tightly by a small cluster of amino acids. Iron, on the other hand, is often trapped in the middle of a large heme molecule, which is held in turn within a tight pocket in the enzyme. Unusual metals, such as molybdenum and vanadium, are used when real force must be applied, as in nitrogenase, the enzyme that separates the two tightly bonded atoms in nitrogen gas.

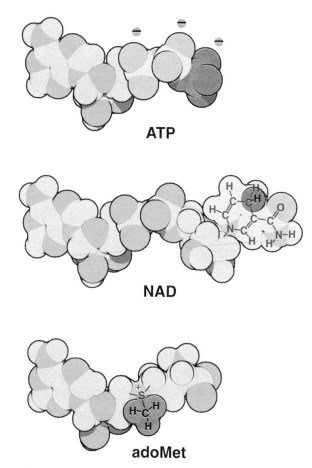

Figure 5-14 Specialized prosthetic groups are used to deliver raw materials to synthetic enzymes. The three shown here are each composed of three parts. At the left in gray is an adenine group, which is used as a handle for holding the prosthetic group in place. This is attached to a specially designed chemical group, shown here in pink, that loosely holds the raw material, shown in red. The phosphate in ATP is easily removed because it is chained uncomfortably close to two neighboring phosphates that carry negative charges. NAD carries a hydride ion (a hydrogen atom with an extra electron). When released, the ring switches to a more stable aromatic structure. Adenosylmethionine (adoMet) carries a methyl group that is held on a sulfur atom. When the methyl group is removed, the sulfur returns to its preferred state, bonded to two atoms instead of three.

REGULATION

To make nanotechnology successful and safe, we must be able to control our nanomachines. We control macroscopic machines with a number of methods. We can control a machine by adding or removing power. This is familiar in appliances that are powered by electricity. Alternatively, we can build a machine that switches between active and inactive states, as in an automobile clutch that switches between a neutral position and a drive position. Or we can physically block the active mechanisms, like a doorstop that blocks the closing of a door. Natural bionanomachines use all of these mechanisms for control.

As described above, several different methods are used for power, including chemical power from ATP, electrochemical gradients, and light energy. In all cases, delivery of the power source may be regulated to control the action of biomolecules. The other two methods, creating a machine with two states and physically blocking the action of a machine, are described below. The molecular analogs to these two processes are allosteric regulation and covalent modification or inhibition. As in our macroscopic machinery, these methods can be combined to form responsive systems of incredible sensitivity and consistency.

Protein Activity May Be Regulated Through Allosteric Motions

Large-scale motions of domains or entire subunits within biomolecular assemblies, termed *allosteric* motions, are widely used for regulation of enzyme activity. Allostery was originally conceived by Monod and coworkers to describe the regulation of binding of small molecules to enzymes. Monod is credited with saying that he had discovered the second secret of life when he discovered this process (the first being information-driven synthesis controlled by DNA). In their definition, an allosteric protein is designed to adopt two (or more) stable structures that can freely interconvert into one another. In one state, the "tense" or T state, the protein resists the binding of substrate molecules. In the other state, the "relaxed" or R state, the protein binds more tightly to substrate molecules. Regulation is accomplished by various methods that force the protein into one of these two states, turning the protein on and off.

In the classic conception, which was developed to describe the unusual oxygen-binding profile of hemoglobin, allosteric proteins show cooperative binding of substrate molecules. They are composed of several identical protein chains, each of which binds to a substrate. Cooperative binding means that binding of one substrate to the protein increases the affinity for binding of additional molecules at the remaining sites. Allosteric proteins use a change of shape to create this cooperative effect. Binding of substrate to the first site is resisted when the complex is in the T state, but once one molecule binds, the entire complex switches to the R state and subsequent substrate molecules bind more readily. Later models showed that this model is somewhat too simple and that the subunits switch from T to R more gradually.

A more general approach to allostery is more widely used. An allosteric complex is built with two different types of binding sites, one for the substrate and another for binding of a regulatory molecule. Binding of regulatory molecules to this second site induces the change in shape that modifies the affinity of the substrate-binding site. These allosteric enzymes have the advantage that no direct interaction between the molecules—between the substrates and the regulatory molecules—need occur. Control is provided completely by the linking of the two sites within the protein complex and is implemented through the change of shape provided by the structural switch. Therefore, the substrates and the regulatory molecules can have very different physical characteristics. For instance, some gene regulatory proteins sense the levels of small molecules such as tryptophan and then change the binding of the protein to entire DNA molecules. These allosteric proteins provide powerful tools for regulating multistep processes, allowing a rich vocabulary of feedback and feedforward regulation of key control points.

Many of the allosteric protein complexes studied at atomic resolution rely on a structural switch (Figure 5-15). The protein adopts two stable states, separated by an energetic transition that is crossed by binding of the regulatory molecule. Many structural features are involved in creating this bistable state. In some cases, the molecule is articulated into several domains, which bend to form the two states. These complexes may be shaped like a pair of scissors that open and close using a flexible hinge. In other cases, the complex is composed of two rings of subunits, which rotate rela-

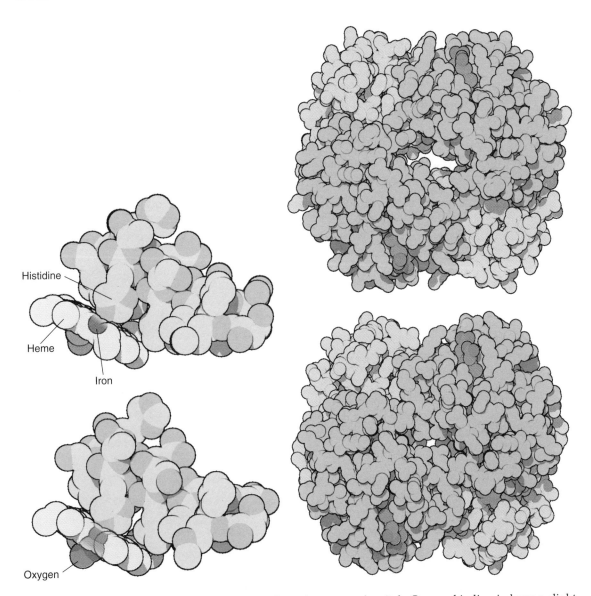

Histidine

Heme

Iron

Oxygen

Figure 5-15   Hemoglobin is regulated with an allosteric structural switch. Oxygen binding induces a slight motion of the iron atom out of the plane of the heme group, as shown on the left. Note that the histidine amino acid is pulled further into the heme ring in the lower oxygen-bound structure. This motion is thought to be the primary signal that powers the allosteric change in shape. Deoxygenated hemoglobin is shown at the top, and oxygenated hemoglobin is shown at the bottom. Note that the subunits move closer to one another in the oxygenated form. The trick used by hemoglobin is the linking of a small motion, which subtly distorts the shape of the local environment, with a sensitive structural switch that flips between these two conformations.

tive to one another to change shape. The two states are stabilized by two different orientations of matching of chemical groups and fitting of knobs into holes.

Very often, the binding sites both for regulatory molecules and for substrates are found between the two subunits (Figure 5-16). The binding sites are composed of amino acids from two neighboring subunits, so when the regulatory molecule binds it can pull or push both of the two neighboring subunits. Also, when the protein changes shape, it can remodel the shape and chemical character of an empty substrate-binding site between subunits, making it more or less accessible for binding of molecules.

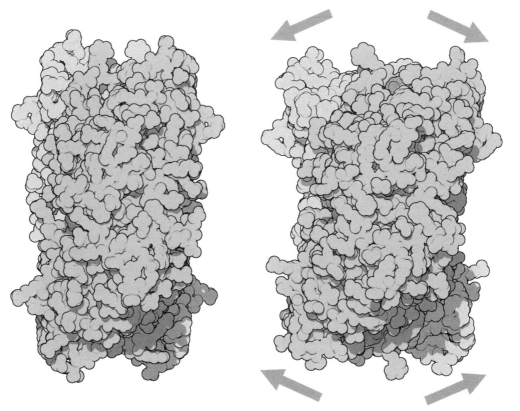

Figure 5-16   The active form of fructose-1,6-bisphosphatase (shown on the left), a flat complex, is composed of four identical subunits. When a regulatory molecule, shown in pink, binds in the space between subunits, it forces a change in shape, scissoring the subunits apart and inactivating the enzyme.

Protein Action May Be Regulated by Covalent Modification

The action of biomolecules is also modified reversibly by covalent modification (Figure 5-17). A chemical group is chemically bonded at one location in the protein. These groups may block the binding of substrates or may cause an allosteric shift in shape that changes the binding characteristics. They may add a new positive or negative charge that changes the interaction with the substrate. Or key chemical groups in the active site may be changed, stopping catalysis.

Phosphorylation is a common approach with many structural advantages. Because much of the natural biomachinery is designed to operate with ATP, many molecular tools are available for adding and removing phosphate groups from various amino acids in proteins. Serine, threonine, and tyrosine are the common targets. Phosphate groups carry a strong charge, so they are easy to recognize and can elicit large structural changes in the modified proteins or interacting partners. Phosphorylation is widely used for passing signals within cells, where messages are received from receptors on the cell surface and carried to sites of action within the cell. Phosphate groups are added by specific kinase enzymes, and the phosphates are removed soon after by phosphatase enzymes to quench the message.

Cleavage of the protein chain is also used widely for regulation in cases where a molecule must be delivered to a site of action and then activated. This is important in the creation of digestive enzymes, which must remain inactive until safely released from the cell. Cleavage is also used to regulate blood clotting, where the components of a clot circulate at high concentrations in the blood, but are only activated when needed at a site of damage. This allows construction of a millimeter-scale structure in a time span of seconds. Cleavage is also used in several aspects of the immune system, where molecules are activated for local defense. In each of these cases, the protein is created in the form of a proenzyme, with an extra segment that blocks the active site or holds the enzyme in an inactive form. Like the pin in a grenade, the blocking segment is removed to create the active enzyme.

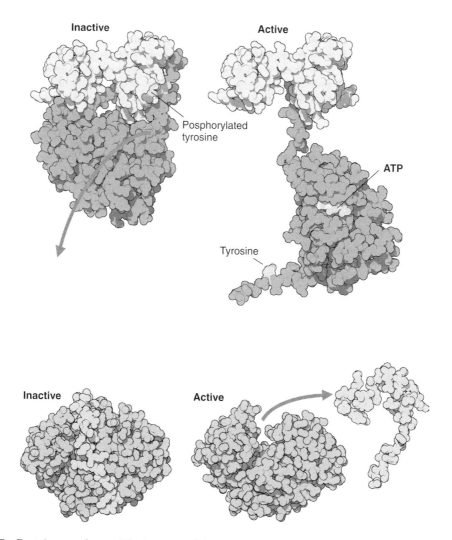

Figure 5-17   Proteins may be modified to control their action. The src protein, shown at the top, uses a phosphate group for regulation. The inactive form has a phosphate attached to a key tyrosine residue, which binds tightly to a small pocket, wrapping the protein in a tight ball. When this phosphate is removed, the protein opens up, revealing the active site, with ATP bound in this structure. Trypsin, shown at the bottom, is activated by cleaving off a regulatory peptide (pink) that normally blocks the deep active site cleft, seen here in profile.

Lessons from Nature

- Bistable biomolecular complexes use allosteric changes in shape for reversible regulation.
- Sites of binding for substrates and regulatory molecules in allosteric proteins are often designed at interfaces between subunits.
- Covalent modification may be used for reversible regulation of molecules.
- Cleavage may be used for one-shot activation of biomolecules.

BIOMATERIALS

Many dreams of nanotechnology center around exotic materials, such as diamond or carbon nanotubes, with exotic properties. Visionaries often design nanomachines or nanorobots that strongly resemble macroscopic machines, shrunk to nanoscale sizes. So naturally they look for tough materials that mimic the strong metal, glass, and plastic materials that we use in our everyday machinery.

Natural biomaterials, on the other hand, are built according to a different paradigm. Organisms are constantly changing, growing, and responding to environmental changes, so biomaterials are dynamic constructions. They are built for a given need and then quickly disassembled when needs change. The structures inside cells may only last for minutes before they are remodeled. Even the most sturdy biological structures, such as bones, undergo continual repairs and reshaping by cells that systematically dissolve and rebuild build them part by part. Remarkably few natural biostructures, such as bone, shells, and wood, remain useful after the signs of life have left; however, they have been used since the very beginning of human technology. We now have the tools to reshape these natural biomaterials and to use the principles of their construction to build our own biologically inspired nanoscale materials.

Helical Assembly of Subunits Forms Filaments and Fibrils

Filaments are created by designing a protein with a binding site for other copies of itself (see Figure 4-15). The use of a self-associating globular protein to form a filament has several attractive properties. Because a filament is composed of a number of modular subunits, it provides a ready scaffolding with many identical attachment sites for other structures. Filaments are also designed for rapid assembly and disassembly, allowing rapid response to the changing needs of a given application. Because the interactions are similar to typical protein recognition sites, the integrity of the filament may be modified by binding of ligands, ions, or other proteins. Helical protein filaments are indeed found throughout nature and are some of the most plentiful and widespread proteins in cells.

By careful design of the location of the self-association sites on each individual subunit, any type of filament may be generated (Figure 5-18). If the sites are on opposite sides of the subunit, the subunits will stack like a string of beads, forming a tenuous, extended filament. If the association sites are shifted slightly, a spring-shaped helix will be formed. Then, if additional self-association sites are engineered, the spring will close down to form a sturdy cylinder. Examples at both ends of this range of design are known.

Actin filaments are an example of an extended filament. Actin is the most common filament-forming protein in our cells, forming the infrastructure of the cytoskeleton and much of the infrastructure used for cell motility. Actin associates to form a directional helical structure with two different ends. When growing, actin monomers add to one end 10 times faster than the other, leading to directional growth of the filament. Actin filaments are highly dynamic in living cells, and are continually built and disassembled minute by minute according to need. Growth of each filament is regulated by the binding of ATP, which promotes growth, and a collection of filament-stabilizing and filament-severing proteins.

Microtubules are an example of a more sturdy cylindrical design. Microtubules are larger, about 25 nm in diameter, with a hollow 14-nm channel down the center. As with actin, microtubules are dynamic structures, with the state of each subunit, free or polymerized, controlled by the state of a bound GTP molecule. They show an interesting property, termed dynamic instability. They lengthen steadily as subunits add to the growing end,

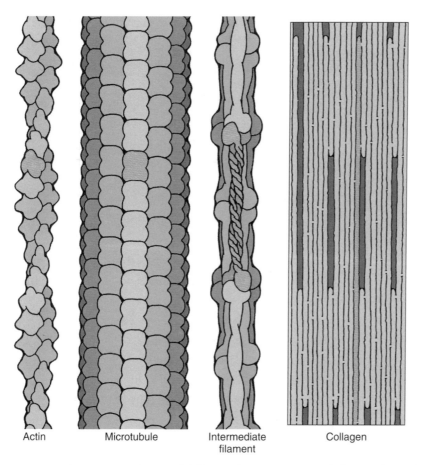

| Actin | Microtubule | Intermediate filament | Collagen |

Figure 5-18 By creating subunits of different shapes, a variety of filaments and fibrils with different characteristics can be formed. Actin filaments and microtubules are continually built and rebuilt and are composed of bricklike subunits. Intermediate filaments and collagen, on the other hand, have extended subunits that overlap extensively and are cross-linked to one another, forming sturdy, ropelike fibrils that resist disassembly. Only a portion of the collagen fibril is shown—see also Figure 2-16.

and then suddenly the growing end frays and the microtubule quickly disassembles. In this way, microtubules are continually extending into new regions but only remain there if specific proteins stabilize the structure. Microtubules play an essential role as tracks for transport within the cell, with objects carried by dynein and kinesin, described below. Perhaps the most spectacular cargo delivered by microtubules are entire chromosomes, which are separated along a beautiful aster of microtubules during cell division.

The dynamic structure of these two filaments combined with their directional nature, with two unique ends, can lead to an unusual behavior termed *treadmilling*. In this behavior, subunits continually dissociate from the lagging end of the filament and reassemble on the opposite, leading edge. Thus the filament stays at the same approximate length but moves through the environment. As described below, treadmilling can be used as part of an engine to power the crawling of cells.

Actin and tubulin are designed for rapid assembly and disassembly. However, this places limits on the strength of the resulting filament. Thin actin filaments are flexible and are easily bent and weak when stretched. Microtubules are also weak when tension is applied but are more resistant to bending. To create structures that are more resistant to tension, a stronger interaction must be designed. In cells, fibrous proteins are used in applications that require strength and durability. These fibrils are also modular, composed of protein subunits, but a combination of lap joints and cross-linking creates a structure that resists disassembly. Two common examples, intermediate filaments and collagen, are found in higher organisms.

Intermediate filaments (so named because they are intermediate in diameter between actin filaments and microtubules) are formed from extended protein monomers that overlap extensively in the filament, strongly interacting along their entire length. The body of the protein is formed from two protein chains that adopt an α-helical structure, with the two long α-helices wound around one another in a strong "coiled coil" structure. Globular domains at each end of the proteins lead to a characteristic knobby appearance of the filaments. Intermediate filaments are used to brace key structures inside cells, forming a loose lattice within the cytoplasm and a tight multilayered sheet just inside the nuclear membrane, termed the nuclear lamina. The similar keratins provide strength to nails and hair. In keratin, added strength is achieved by linking cysteine amino acids between proteins, forming strong covalent disulfide bonds.

Collagen is the most common protein in the human body. It is the primary fibril in structures that support and connect cells into tissues. In its simplest form, collagen is a long, thin protein composed of three similar protein chains that wind around one another in a characteristic triple-stranded helix. This tight stable structure requires a special repeating amino acid sequence, of the form glycine-X-proline, where X can be any amino acid (see Figure 2-4A). The glycine is needed to fit inside the tight triple helix at a given point each turn, and the proline is needed to bend the chain back tightly enough to continue the helix. In addition, many of the proline amino acids are modified with an extra hydroxyl group, allowing them to form additional hydrogen bonds with neighboring strands. These long proteins then associate side by side in staggered registrations to form collagen fibrils. The proteins are cross-linked to neighbors through lysine-lysine covalent bonds for added strength.

Microscale Infrastructure Is Built from Fibrous Components

When building microscale and larger structures, nanoscale structural units must be combined to give bulk biomaterials. We might imagine building a solid structure, like a brick building, with modular protein subunits. However, this approach is only rarely used in natural structures. Most biological infrastructure is created as an extended network of firmly linked nanoscale components. These may take the form of two-dimensional fishnet structures, networks that fill three dimensions, or variations between these extremes. These networks are strong but resilient at the same time. They are typically porous and allow free transit of water and small molecules.

One of the most widespread approaches is to create a tough two-dimensional network to brace lipid membranes. Lipid bilayers have excellent permeability properties but cannot withstand many of the pressures imposed by the environment and by differences in osmolarity. Specialized networks of polymeric molecules are used to brace cell membranes directly, providing support.

Bacterial cells use a network of peptidoglycan to brace their cell walls. Peptidoglycan is composed of long carbohydrate chains cross-linked by short peptide strands. A collection of assembly enzymes is required for assembly of the structure (penicillin works by attacking one of these enzymes). The simple fishnet form of peptidoglycan allows the bacterium to perform a particularly tricky construction feat when dividing. The cell is under high osmotic pressure, because of the high concentration of molecules inside, and so the peptidoglycan layer must be kept intact during the entire process of lengthening the cell. This is thought to be accomplished by creation of new carbohydrate chains parallel to two existing cross-linked chains. The new chain is then cross-linked to the two existing chains, and then the cross-links between the existing chains are severed, allowing the network to expand.

Inside animal cells, a network of proteins is used to brace the inside of the cell membrane. Anchoring proteins that span the membrane are connected to actin filaments that extend into the cell, forming part of the cytoskeleton. A geodesic net of spectrin, a long two-chain protein, then links the actin filaments just below the surface. Red blood cells have a highly simplified version of this network, with just a short segment of actin providing the tether for the spectrin net (Figure 5-19). These cells are highly flexible and are able to bend almost in half, but the spectrin net maintains a constant surface area as the cells push through narrow regions of the circulatory system.

A three-dimensional network of actin filaments, along with intermediate filaments and microtubules, creates the internal infrastructure used for support and as the roadways for transportation (Figure 5-20). Actin filaments are linked together by actin-binding proteins to build the cell-spanning microscale structure. Actin filaments by themselves form a viscous liquid, but if the protein filamin is then added, it cross-links the actin filaments into a loose three-dimensional network that is a semisolid gel. Cross-linking proteins with different shapes are used to create networks for different functions. Filamin is shaped like a flexible hinge, so it forms loose tangled networks. Fimbrin, on the other hand, is a small, rigid protein with two actin-binding sites. It aligns actin filaments side by side and is used to create rigid bundles. These are used in some cells to support fingerlike structures that extend from the surface of the cell.

As we move from microscale to macroscale structures and look at organisms that are built of multiple cells, these problems magnify. Additional levels of bracing are necessary. Many of the structures bracing and supporting larger organisms use the approach of reinforced concrete. A stiff linear fibril, analogous to steel reinforcing rods, is combined with a less sturdy

Figure 5-19 The cell membranes of animal cells, shown here in light pink, are braced by a geodesic network of the protein spectrin, shown in dark pink. A cross section through a red blood cell membrane is shown here (with lots of hemoglobin inside at the bottom).

space-filling matrix, analogous to concrete. Together they create a composite with the best features from both.

Basement membranes are tough, sheetlike structures that are used in higher organisms between cells to support tissues (Figure 5-21). They also act as molecular sieves that block the passage of large molecules but allow passage of small molecules. Basement membranes combine a tough fibrous

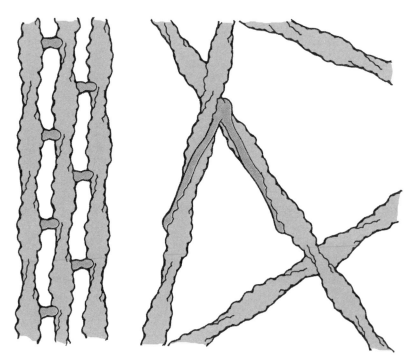

Figure 5-20   Actin-binding proteins glue actin into different infrastructures. On the left, many small fimbrin proteins have bundled actin filaments in parallel. On the right, large hinge-shaped filamin arranges actin into a looser three-dimensional network.

network with a carbohydrate-rich gel matrix. The underlying network is composed of two interconnected networks of collagen and laminin. Collagen (type IV) forms long cables that associate through their free ends to form an extended polyhedral network. Similarly, laminin, a cross-shaped protein complex of three chains with a number of sticky ends, forms a network though association of its ends. These two networks interpenetrate and are connected by the protein entactin. This loose but strong network is filled with heparin sulfate glycoproteins, protein complexes bristling with many carbohydrate chains. The carbohydrate chains contain sulfate groups that carry a strong negative charge. They bind strongly to laminin, gluing the proteoglycans into the collagen/laminin network.

Plants use a similar approach in their cell walls. The fibrous component is composed of cellulose, a carbohydrate polymer composed of several

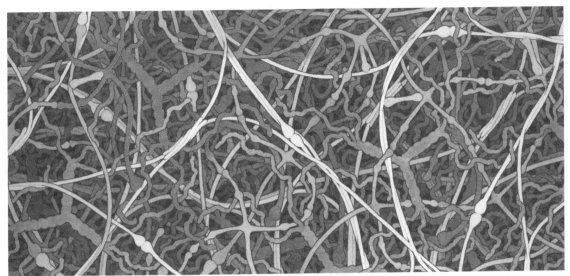

Figure 5-21 Basement membrane is a composite structure composed of long, thin collagen cables, cross-shaped laminin molecules, and a space-filling matrix of three-armed proteoglycans.

thousand glucose subunits. The chains of cellulose associate in parallel to form strong microfibrils. Hydrogen bonds between the neighboring strands are so well ordered that the cellulose strands form a crystalline structure inside these microfibrils. A variety of hemicellulose molecules, shorter carbohydrate polymers, link the cellulose microfibrils by forming hydrogen bonds with the molecules on the surface. The interstices are then filled with pectin molecules. Pectins are linear or branched carbohydrate chains with a high proportion of d-galacturonic acid, a negatively charged sugar. These interact strongly with calcium ions, forming a gellike network of linked chains. Together, the entire structure is strong enough to brace the tallest living things.

Minerals Are Combined with Biomaterials for Special Applications

When additional strength or altered properties are needed, minerals are added to biomaterials. More than 60 different types of minerals have been discovered in biological systems. In bones, teeth, eggshells, and seashells, minerals are incorporated to add strength, but crystals also play less familiar functional roles. For instance, the inertia of small otolith crystals is used to sense gravity, and the alignment of small of magnetite crystals is used to sense the magnetic field of Earth. The optical properties of single crystals are also exploited, albeit more rarely than in our commercial world: Trilobites used crystals of calcite in their eyes, and some scarab beetles owe their metallic luster to crystals of uric acid.

Biomineralization is a fascinating process that is just starting to reveal its secrets (Figure 5-22). The growth of crystals on demand requires control of four processes. First, a space within the biological matrix is created to allow growth of the mineral. Second, ions are transported into the space, often at very high concentrations. Third, crystals or aggregates of the mineral are nucleated at the desired locations. And finally, the growth and orientation of the mineral must be carefully controlled to produce the desired size and shape. By using different approaches to these steps, different types of minerals are tailored for specific applications.

A common approach is to create a lipid-bounded vesicle and then to pump a high concentration of ions inside, allowing them to nucleate at many sites. This is the fastest method of biomineralization, and it results in the formation of a porous mass of random spherulites, similar to a ceramic. If specific nucleation sites are added, a more ordered biomineral can be formed. In the eggshells of birds, sulfated polysaccharides are found on the inner surfaces of the mineral-producing vesicles, which nucleate the crystals and align them as they grow outward from the wall.

Sea urchins show even more control. As they create their spiny shells, they create a structure composed in large part from a single crystal of calcite. A large vesicle is created by fusing many cells together. The single crystal is then nucleated and its growth carefully controlled, presumably through use of specific proteins that alter the growth rate along different crystal faces. The result is a plate or spine up to a centimeter in size composed of a single crystal. Similarly, our bones are created by the nucleation and growth of single crystals of dahllite inside a matrix of collagen.

Incorporation of biological materials with minerals can strengthen the composite. These sturdy organic-inorganic composites are among the most exciting potential raw materials for use in bionanotechnology. They combine the strength of inorganic materials with the resilience of biological ma-

Figure 5-22 Examples of biomineralization. A. Aragonite spherulites from a chiton are randomly oriented and loosely packed. B. Calcite crystals from a hen's egg are nucleated at the inner surface of the shell, seen here at the bottom of the picture, and grow upward, fusing into columnar structures. C. Calcite crystals from a bivalve shell are each nucleated separately within an organic matrix, forming perfect prisms. D. A mineralized collagen fibril from turkey leg tendon shows many plate-like dahllite crystals (the dark lines are crystals seen on edge). (Figure 7 from Addadi, L. and Weiner, S. (1992) "Control and Design Principles in Biological Mineralization." *Angew. Chem. Int. Ed. Engl.* 31, p. 166.)

terials. For instance, the calcite crystals used in sea urchin spines are prone to fractures, so sea urchins incorporate about 0.2% by weight of a small peptide, which reinforces the structure and prevents fractures. The combination of collagen with mineral makes bones strong but resilient, resistant to fracture. The pearly coating of oysters and other mollusk shells is a more elaborate hybrid material. It is composed of alternating layers of biomolecules and minerals. The biomolecule layers are themselves composed of separate layers: chitin (a polysaccharide) at the center, with layers of a silklike protein on each side, and then a layer of acidic proteins on the two surfaces. The acidic proteins nucleate and control the growth of aragonite crystals, which align between the biomaterial layers. The many aligned layers of crystals provide the interference of light that gives these shells their iridescent luster.

The molecular mechanisms of nucleation and growth are now becoming better understood, and researchers are applying them to bionanotechnology. Minerals composed of silica, such as those in sponges and diatoms, appear to be constructed by proteins with abundant serine or cysteine amino acids. These amino acids perform the catalytic task of connecting many silica molecules into an amorphous glass. Minerals composed of calcium ions, on the other hand, appear to be constructed by polymers of acidic amino acids. Surprisingly, these can be used in two exactly opposite roles depending on their conformation. Free chains of acidic amino acids associate with the surfaces of growing crystals, blocking growth and providing a mechanism for shaping minerals. If, however, these chains are immobilized on a surface, aligning the acid groups properly, they can nucleate new crystals. Globular proteins can play the same role, aligning acid groups on their surfaces with the proper orientation for crystal nucleation.

Elastic Proteins Use Disordered Chains

Elastic materials may be stretched and distorted, and then when released they snap back to their original shape. At the molecular level, elastic materials are composed of many randomly coiled chains. When stretched, the coils unwind, reducing the high level of contact that is formed inside the coils. When released, the chains collapse back into the densely coiled form. As long as the chains do not slide relative to one another, the material will return to its original shape.

Natural rubber, which is obtained from the milky latex secreted by rubber trees and other plants, is formed of long polymers of isoprene. Because these hydrocarbon chains simply lie next to one another, some slippage between chains is possible, so natural rubber does not resume its shape perfectly after stretching. However, in 1839, Charles Goodyear discovered a way to stop the slippage, termed vulcanization. The rubber is heated with sulfur, which forms disulfide bridges at many points between chains. This cross-linked network is very resistant to slipping, so vulcanized rubber can undergo many rounds of stretching without losing its shape.

The protein *elastin*, which provides much of the elasticity in skin, uses a principle similar to that of vulcanized rubber (Figure 5-23). It is composed of sections that are rich in proline amino acids, which form random coils that extend when stretched. Between these random coils, there are short segments that contain lysine amino acids. The side chains of these lysines form cross-links to neighboring chains, forming an elastic network.

The giant protein titin, which gives muscle tissue elasticity, uses two different mechanisms. Titin is a long protein composed of many individually folded domains arranged like a string of beads. At one end is a special region, termed the PEVK region because of the abundance of proline, glutamic acid, valine, and lysine, which provides elasticity at normal levels of

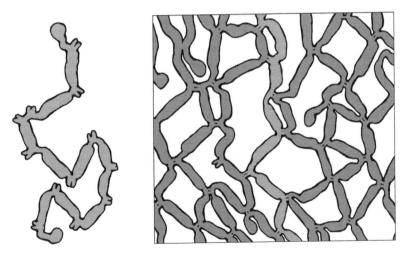

Figure 5-23 Elastin is composed of a series of protein segments with little structure, shown here as sausage shapes, with many lysine amino acids in between the segments, shown here as the little pairs of knobs extending from the chain. The lysines cross-link with neighboring chains. The result is an elastic network, shown on the right, that stretches when pulled and snaps back to this compact structure when released.

force. When relaxed, it forms a loose, random globule, with the chain compacted but with no stable folded structure. As the molecule is pulled, this globule slowly unfolds, resisting as the interacting parts of the chain are separated. When released, it collapses back into the loose globule. But when titin is pulled even harder, the string of stable domains unfold, one after the other. The unfolding is reversible, so they can snap back into their shortened, folded form when released. These little 20-nm steps, as each domain unfolds, have been observed when titin is stretched with optical tweezers or atomic force microscopes.

A similar, but less permanent, structure can be used to create a hydrogel. We are all familiar with the formation of hydrogels by the natural protein gelatin. A water solution of gelatin is heated, and as it cools it traps water to form a flexible gel. The process is reversible. The gel may be resolubilized by heating and resolidified by cooling. The molecular mechanisms of gelation are poorly understood, but researchers are attempting to design and enhance molecules with these properties. Like elastic materials, gels are formed of long, disordered chains that are connected at specific points. But the connections must be temporary. To make a useful gel, the interaction points between the chains must be strong in the gel but must release in when the gel melts. The disordered regions must also be carefully tailored. The chains must remain accessible to solvent as the gel forms, or the chain will precipitate as the solution cools. So they cannot form ordered structures like the structures in folded protein chains.

David A. Tirrell and coworkers have developed an artificial protein system to test this strategy for design of hydrogels. Their proteins are composed of two parts. At each end, there is a protein sequence that forms an α-helix at room temperature, with a line of leucine amino acids arrayed on one side. These associate side by side to form a *leucine zipper* that links two chains together. Leucine zippers were first discovered in bacterial DNA-binding proteins. They bind tightly once the two halves meet but are released when heated. Between these two elements, they have put a long chain that is rich in glycine, proline, and glutamic acid. This portion forms a long, disordered structure at a large range of temperatures. As hoped, this protein forms gels when heated and then cooled. The modularity of the system is attractive and will allow a variety of proteins to be designed with different gelation properties simply by changing the length and character of the different modules.

Cells Make Specific and General Adhesives

Adhesives are a remarkably cost-effective and general method of construction. If designed correctly, adhesives can be used to join many dissimilar materials with a stress-resistant bond. Adhesives require two design criteria: They must form a strong, intimate interaction with the surfaces being connected, and they must be cured into a tough solid, so that the join itself is stable. In our water-filled world, however, adhesives are often compromised by the presence of water or water vapor. Water can form a thin layer on surfaces, blocking the interaction of adhesive with surface. Water can also attack already-glued areas, destroying the glue or infiltrating along the seam between adhesive and substrate.

Natural systems use two different general methods for adhesion. Most of the adhesive mechanisms that hold together your body, gluing the cells and the extracellular infrastructure together, are the result of specific adhesion. This is mediated directly through specific protein-protein and protein-carbohydrate interactions. A host of proteins form specific interactions that glue cells together. These include junction proteins such as claudins that seal two cells tightly together, connexon proteins that connect cells with a intercellular pore, and cadherins that extend from the cell surface and associate in the presence of calcium ions (Figure 5-24). Integrin proteins extend from cell surfaces and attach to the infrastructure between cells. Because all of these proteins use the powerful methods of protein recognition, the interactions are strong. These modes of adhesion are also very specific. They only work if the proper partners are present on the two structures to be connected and thus can be used to create adhesive contacts exactly when and where they are needed.

In special cases, cells have developed general adhesives that work on a variety of surfaces. A well-studied example is a class of adhesives used by marine shellfish to attach themselves to rocks. These adhesives are formed in seawater and are very strong, gluing the animals to rocks against the forces of pounding surf. A class of specialized proteins forms the adhesive.

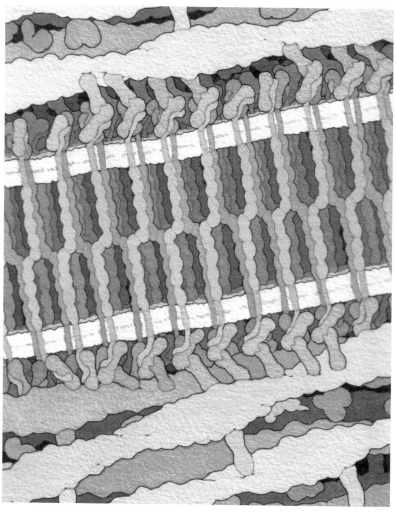

Figure 5-24   Cadherin, shown here in pink, forms specific contacts between cells, forming an adhesive that holds cells together.

These proteins contain many amino acids that have extra hydroxyl groups. These serve two purposes. First, because hydroxyl groups act both as hydrogen bond donors and acceptors, and because they interact strongly with metal ions, they are excellent for forming multiple interactions with the surfaces being bonded. Second, these modified amino acids are used to create many cross-links when the adhesive is cured, changing the adhesive from a

---

Lessons from Nature

- Filaments are created by helical association of protein subunits.
- Fibrils may be strengthened by increasing the overlapping surfaces of subunits and by cross-linking subunits.
- Strong, resilient composite materials are composed of a two-dimensional network impregnated with a space-filling matrix of molecules.
- Biominerals may be grown in closed spaces by raising the concentration of ions and providing specific nucleation proteins.
- Elastic proteins contain disordered segments and segments that cross-link.
- Hydrogels are built from proteins that contain disordered segments and segments that form reversible interactions.
- Marine bioadhesives use hydroxylated amino acids that interact with surfaces and allow extensive cross-linking during curing.

thick liquid into a tough resin. These remarkable adhesives are being explored for use in dentistry and medical applications, where adhesives must be resistant to water.

BIOMOLECULAR MOTORS

When we think of machines, we think of moving parts powered by some type of motor. Quite naturally, many of the speculative designs from molecular nanotechnology mimic our macroscale machines, with turning wheels and gears and nanoscale motors to turn them. Powered motion is appealing, because it provides a level of direct control that is not available with other methods.

In bionanomachinery, however, motors are remarkably rare. Much of the work of nanoscale transport and motion is accomplished by diffusion and capture, without the need for directed motion. Diffusion is so fast for proteins in cell-sized enclosures that no additional mechanisms are necessary. Motors are brought to bear in larger tasks of motion at the micrometer level, such as the separation of chromosomes and the remodeling of cell organelles, and all the way up to meter scales, with the contraction of our muscles. However, there are a few remarkable exceptions where nature does use motors for nanoscale tasks, such as the rotary motor in ATP synthase.

ATP Powers Linear Motors

Looking in cells, we find that several different approaches are used to power linear motion along a fixed track. The two best-studied examples are *myosin*, which moves along actin filaments, and *kinesin*, which moves along microtubules. These two motor proteins are quite different. Each myosin molecule performs one "power stroke" at a time: It binds to actin, pulls on it, and then dissociates. So, to do any real work, we need to have a lot of myosin molecules working in concert. Kinesin, on the other hand, is highly processive, attaching to microtubules and making many successive steps before separating. These mechanistic differences suit their biological functions: Myosin is found in large arrays for moving long actin filaments—each myosin must perform its power stroke and then get out of the way to allow the neighboring molecules to do their jobs. Kinesin acts more autonomously. It is used like a locomotive to pull cargo along microtubules. It must attach and then step continuously down the microtubule track. Despite these different approaches, myosin and kinesin show many similarities in the molecular machinery used for force generation.

The motor domain of myosin is elongated, with a site for binding of ATP near the center (Figures 5-25 and 5-26). Myosin is normally bent, and the ATP-driven power stroke causes the molecule to straighten. The motor domain is composed of several functional parts. A large *catalytic* domain at one end binds to ATP and to actin. ATP-driven motions of the catalytic domain cause a change in the *converter* domain, which are then magnified by the *lever arm*, producing a 100-Å displacement at the end. In different forms of myosin, the tips of these lever arms are connected to appropriate handles for each job.

Kinesin uses an entirely different approach to motion (Figure 5-27). It

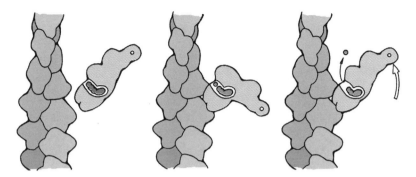

Figure 5-25 Myosin cycles through three states to provide directional powered motion. The first step, shown on the left, has ATP bound, and it does not bind to actin. In the second step, shown in the center, the ATP has been cleaved to ADP and phosphate. This causes a shift in the actin-binding face, allowing it to bind strongly to the actin filament, and cocks the lever arm into a bent state. The phosphate dissociates in the third step, causing the myosin to straighten, performing the power stroke. The remaining ADP will then be replaced by a new ATP, dissociating the myosin from the actin filament and making it ready for the next stroke.

uses an unusual order/disorder transition in a short stretch of the protein chain, termed the *neck linker*. Powered by ATP cleavage, the neck linker changes from a disordered form into an ordered form to create the power stroke. The kinesin motor domain has a groove that binds transiently to this chain and releases it on command. The power stroke is performed when this linker segment zippers into the channel. Anything attached to the linker is dragged along, moving it by 80 Å.

These motors sense the presence of a single phosphate group on ATP and use cleavage of ATP to create a powered structural change. In both myosin and kinesin, two sensor loops are used to translate ATP binding into a structural change. They form hydrogen bonds with the phosphate, acting like a spring-loaded gate that closes around the group and opens when the phosphate is cleaved and released. The key structural transitions in both cycles occur when ATP binds, which closes the switch loops, and when phosphate is released, which reopens the switch loops. In myosin, ATP binding causes a significant structural change from a rigidly extended form to the bent form and phosphate release allows the opposite transition. In kinesin, ATP binding allows the zippering of the neck linker into the

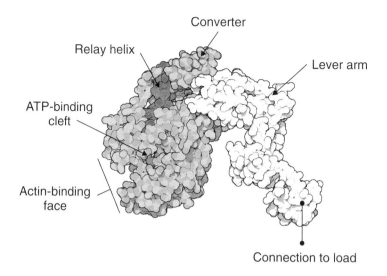

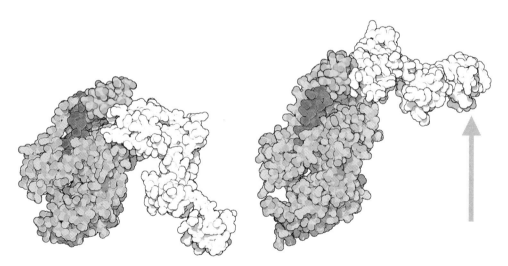

Figure 5-26 Crystallographic structures have revealed the atomic details of the myosin power stroke. Myosin binds ATP in a deep cleft, as shown at the top, with the actin-binding face on one side of the cleft and the power stroke machinery on the other side. The relay helix acts like a piston to transmit the small structural changes as ATP is cleaved and phosphate is released. This piston then pushes on the converter, which causes the large swinging motion of the lever arm, as shown at the bottom. The cocked state is shown on the left, with the relay helix pushed upward into the converter, and the state after the power stroke is shown on the right, with the relay helix pulled downward.

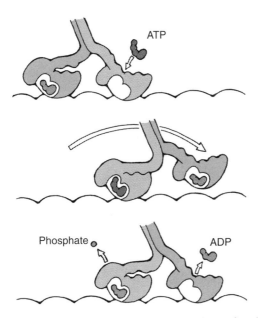

Figure 5-27 Kinesin relies on two motor units connected together. The cycle begins with one subunit empty and the other with ADP bound. ATP binds to the empty subunit and causes the neck linker to zip tightly onto the subunit, pulling the lagging subunit off the microtubule and forward to the next position. When it binds, ADP is released. Cleavage of the ATP and release of phosphate in the new lagging subunit releases the neck linker, allowing it to take its unbound, disordered form and readying the complex for the next step. Successive cycles allow kinesin to walk along the microtubule.

groove and phosphate release allows the linker to dissociate into its disordered, unbound form (Figure 5-28).

The conformational change of the switch loops is transferred to the motility elements by a long *relay helix*. This structure is a stable α-helix and thus is rigid and incompressible. It acts as a piston to transmit the motion of the loops. The helix moves toward ATP as the switch loops close around it, and the helix moves away when phosphate is released and the loops open. In myosin, the relay helix contacts the converter region, causing it to rotate like a hinge around two flexible glycine amino acids. This rotation is transferred to the lever arm, which is rigidly attached to the converter. In kinesin, motion of the relay helix opens a new pocket, allowing the binding of the neck linker. Again, a glycine residue is used in the neck linker to allow

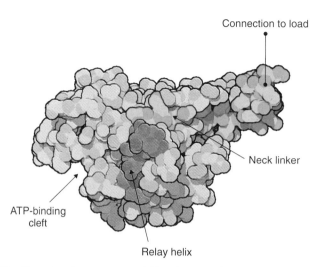

Connection to load

Neck linker

ATP-binding
cleft

Relay helix

Figure 5-28 The atomic structure of kinesin reveals machinery that is similar to myosin. The surface that binds to the microtubule is along the bottom in this view. When ATP binds, it shifts the position of the relay helix, which creates the long, narrow groove that holds the neck linker. Force is generated when the neck linker zips tightly into this groove, as seen in this structure.

the flexibility needed to bind and release from the pocket controlled by the relay helix.

Quite remarkably, these motors may be isolated and used to perform custom tasks. For instance, kinesin may be isolated and bound onto a flat surface of a microscope slide. Then, if microtubules are added to the solution, they will be pushed around on top of this bristling array of kinesin, like a molecular mosh pit.

ATP Synthase and Flagellar Motors Are Rotary Motors

Nature has also developed nanoscale rotary motors. The most familiar example, which has already been harnessed for bionanotechnology applications, is ATP synthase. ATP synthase is a combination of two motors, termed F_0 and F_1, which are powered by two different fuels (Figure 5-29). The F_0 motor is powered by a proton electrochemical gradient, and the F_1 motor is powered by ATP. In the complex, the two motors are connected so that it acts as both a motor and a generator. Motion of F_0 driven by proton flow can be used to generate ATP in F_1. Alternatively, by rotating in the op-

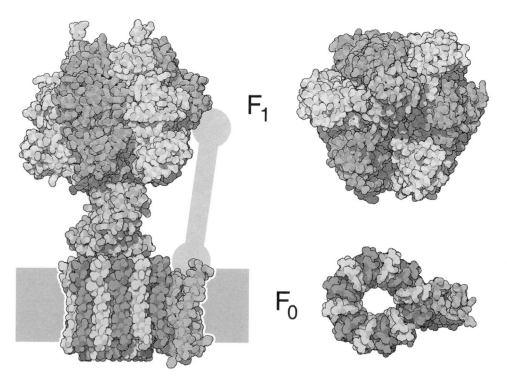

Figure 5-29 ATP synthase is composed of two tethered nanomolecular motors. The F_0 motor at the bottom is embedded in a membrane and is composed of a rotor, shown in gray, and a stator subunit, shown in pink. An eccentric axle extends up from the rotor and passes through the center of the F_1 motor, distorting the six subunits in F_1 as it turns. The large arm connecting the F_0 rotor to F_1, shown schematically here in pink, has been seen by electron microscopy but not in atomic detail.

posite direction, ATP-powered motion of F_1 can be used to turn F_0 and pump protons, creating an electrochemical gradient.

The F_0 motor is composed of a cylindrical rotor, with 10–14 identical subunits (depending on species) and a stator that associates with one side of the rotor (Figure 5-30). The entire complex must be embedded in a lipid bilayer membrane, which provides the barrier across which protons flow and is essential for the mechanism of transfer. Two forms of the motor have been studied extensively, one powered by protons and one powered by sodium ions, and several different models have been proposed for their action. The most popular model for action relies on rotational diffusion, which is biased to turn in one direction by the flow of protons or ions through the system.

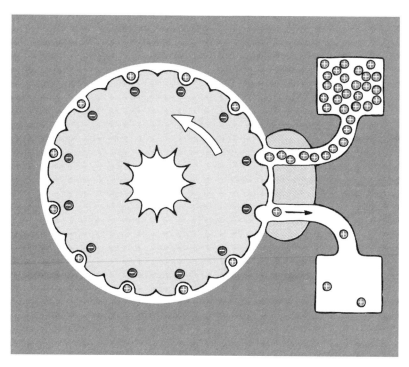

Figure 5-30   The F_0 rotor of ATP synthase has a binding site for protons that carries a negative charge. Because it is buried in the membrane, it can only turn if the charge is neutralized by a proton. The stator, shown in pink, supplies the protons from one side of the membrane and deposits them on the other side.

The rotor will naturally perform a rotational random walk, making small random steps forward and backward under the power of thermal motions. The trick is to favor steps in the forward direction and to block steps in the reverse direction. The proton-driven F_0 motor contains an acidic aspartate amino acid on the surface of the cylinder, buried in the middle of the lipid bilayer. At typical pH, aspartate carries a negative charge, which resists contact with the membrane. The F_0 motor is thought to use this resistance to power the motion. The stator protein covers one side of the rotor and contains two channels, one that leads from the side with a high concentration of protons and one that empties into the other side that is low in protons. In this shielded area, the aspartate is free to be charged, because it is not in contact with the membrane. Protons enter through the stator from the side with the high concentration and associate with the aspartate, neutralizing its charge. The rotor is then free to rotate, exposing the neutralized aspartate to the membrane. As the rotor turns, the protonated aspartate travels all the way around, until it enters the stator-covered region again. There, it encounters the exit path, and the proton dissociates and diffuses away into the region of low concentration, leaving a charged aspartate that cannot rotate backward.

The entire process is reversible. If left alone, it will pick up protons on the side of high concentration, turn all the way around, and release them through the channel to the side with low concentration. If the rotor is driven the other way, however, it can pump protons. A constant rotational force is applied, but it cannot turn until it picks up a proton on the low-concentration side. Then the rotor turns all the way around, and the proton is released at the high-concentration side.

This F_0 motor, driven by protons or sodium ions, is connected to the F_1 motor through an axle. The F_1 motor is driven by the cleavage of ATP. F_1 is composed of a ring of six subunits, three α-subunits and three β-subunits. An eccentric axle runs through the center of the ring, pressing differently on the three β-subunits and markedly changing their structure. (The α-subunits are apparently necessary for the structural integrity of the whole complex but do not participate directly in the rotary mechanism) The different conformations of the β-subunit have markedly different affinity for ATP. Early work by Paul Boyer identified three forms, labeled tight, loose, and open, and recent work has added more detail. Figure 5-31 shows a model based on direct observation of rotation in a system in which a bead was attached to the axle and observed by microscopy. Each rotation of 120° occurs in two steps. When rotation is powered by ATP, a 90° rotation is powered by binding of ATP. The ATP then hydrolyzes into ADP and phosphate with no change in conformation. The dissociation of ADP then fuels a 30° rotation.

The process works in both directions. Cleavage of ATP can cause the rotor to turn, or forcible rotation of the axle can distort the three β-subunits, causing them to adopt conformations that would not normally be favorable (Figure 5-32). ATP synthase is then an ATP generator. The forcible rotation of the axle first increases the affinity for ADP, which combines with phosphate to form ATP without change in conformation, and further rotation forces ATP to be released.

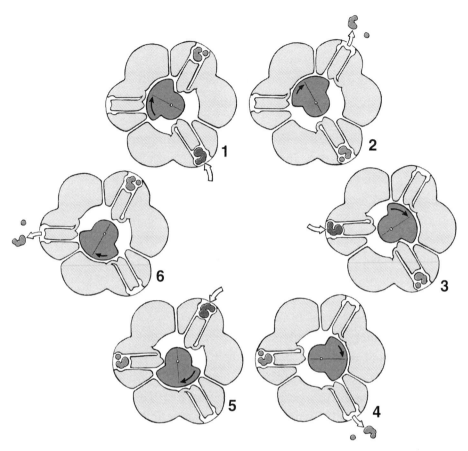

Figure 5-31   The rotary cycle of ATP synthase has includes two types of rotary steps. In step one, ATP binds, causing a 90° rotation. In the second step, ATP and phosphate from an adjacent site leave, causing an additional 30° rotation. By repetition of these steps three times, ATP synthase makes a full revolution.

Note that the F_1 motor is thus not directly fueled by ATP cleavage. ATP cleavage instead provides an essentially irreversible step to the cycle. When acting as a motor, ATP is easy to bind to the motor but difficult to release and ADP is difficult to bind but easy to release. ATP binds easily, is converted, and leaves easily as ADP. In the synthetic direction, the opposite approach is taken, and the powered motion is used to assist the two difficult steps. ADP binding is assisted, and ATP release is assisted.

Many bacteria build a much larger rotary motor for use in propulsion (Figure 5-33). These bacteria swim by rotation of a long corkscrew-shaped

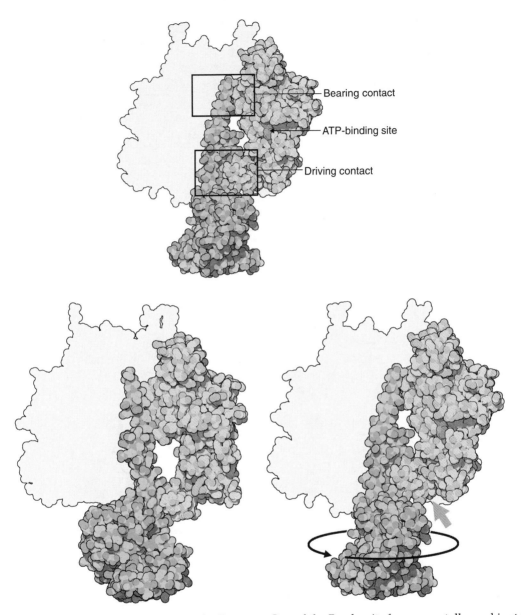

Bearing contact

ATP-binding site

Driving contact

Figure 5-32 Conformational changes in the F_1 motor. One of the B-subunits from a crystallographic structure of the F_1 motor is shown at the top in pink and the bound nucleotide in darker pink. The central axle is shown in gray. At the upper end, the B-subunits form a bearing that contacts the axle but does not change conformation as the axle turns. Two positions of the axle are shown at the bottom. Note that the upper half of each subunit is in a very similar position. The axle moves the lower half, pushing it away from the central axis strongly in the form that does not have nucleotide bound. This changes the character of the active site, which is formed between the bearing domain and the mobile domain.

Figure 5-33 The flagellar motor of *Escherichia coli* spans the two-layered cell wall of the bacterium and turns the long corkscrew-shaped flagellum. The other rotary motor of the cell, ATP synthase, is also found spanning the cell wall, shown in darker pink in this illustration.

flagellum powered by a switchable motor that can power rotation in either direction. Bacterial flagellar motors are spectacular examples of rotary motion, rotating at rates of over 100,000 rpm and driving cells at speeds exceeding hundreds of micrometers per second. They provide a constant torque at a wide range of speeds.

As with the F_0 motor of ATP synthase, the bacterial flagellar motor is composed of a membrane-bound stator and a membrane-spanning rotor. The details are still under intense study, but the basic morphology and composition of the motor are known. There appear to be a series of stator complexes arranged in a ring about the rotor complex, which is composed of multiple copies of several proteins with higher symmetry. There appear to be eight torque generating units, created by the combination of stator and rotor units. Approximately 400 force-generating steps provide one rotation and require transfer of about 1200 protons across the membrane. In other bacteria, a gradient of sodium ions is used for power instead of a proton gradient.

Brownian Ratchets Rectify Random Thermal Motions

The nanoscale world is dominated by thermal motions, and, not surprisingly, cells have developed machines that capture thermal energy and use it to do work. The trick is to create a machine that rectifies random thermal motion. For instance, imagine a particle diffusing along one dimension. It performs a random walk along a line. Now, place a barrier that allows passage in one direction but is impassible in the other. Random thermal motion will eventually push the particle through the barrier but will be unable to return it to the other side. In a system with many particles, particles will be pumped to one side of the barrier, using thermal energy to cross the barrier.

This principle may be extended to create structures that apply force, termed *Brownian ratchets*. We have seen one example above: the F_0 motor of ATP synthase. In that motor, the rotor undergoes random rotational fluctuations, which are rectified by the passage of protons through a barrier. A difference in concentration between the two sides ensures that protons pass predominantly in one direction.

The best-characterized example of a Brownian ratchet is actin, which is used in the crawling of cells (Figure 5-34). Actin polymerizes into long fila-

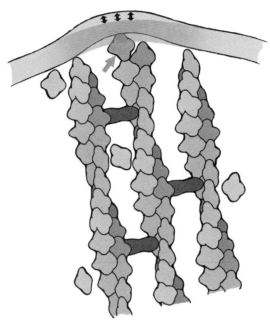

Figure 5-34 Actin acts as a Brownian ratchet to extend membranes. The membrane, shown at the top in pink, undergoes random thermal fluctuations, which transiently open up enough room to add another actin subunit to the growing filament. Cleavage of ATP in the newly added actin subunit glues it in place, holding the membrane in the extended position.

ments, using ATP to drive the transition from free subunits to assembled polymer. The Brownian ratchet is created by fixing the tail end of the actin and placing the growing end against the load, in this case, a membrane. Random motions will displace the membrane, occasionally allowing room for a new actin monomer to add. When the filament extends, the membrane is held in its new position and performs a new random fluctuation around that position. Again, actin monomers may add when the fluctuations allow more room. The ratchet is created from the combination of random fluctuation of the load and irreversible assembly of the filament applying force. The one-way barrier is provided by the cleavage of ATP, which locks each actin in place.

In cells, actin is used to create the force for moving the cell. A particularly parsimonious approach is taken through the process of *treadmilling*. Actin filaments are extended by removing subunits from the tail end, allow-

---

Lessons from Nature

- Power strokes of ATP-fueled molecular motors are powered by the binding of ATP and/or the release of ADP and phosphate. The cleavage reaction provides an irreversible step that makes the process directional.
- Multi-nanometer scale motions can be powered by linking an ATP-cleavage site to a protein conformational change. Examples include a series of articulated motions, as in myosin and ATP synthase F_1, or motions that drive specific order/disorder transitions, as in kinesin.
- Thermal motion can be rectified by a Brownian ratchet. These require one-way barriers to provide rectification. Examples include the charge-neutralization gate used in the ATP synthase F_0 motor and ATP used in actin polymerization.

---

ing them to diffuse forward, and then adding them to the leading end. This process occurs naturally in actin solutions, but is accelerated up to 100 times in cells in several ways. Two proteins modify the actions at each end: Actin depolymerizing factor aids in the breakdown of the tail end, and profilin modifies actin into the conformation that is best for addition to the leading edge. Other tricks are also used to enhance growth rates. For instance, many actin filaments may be present at a surface being moved. Many of these may be capped with proteins that inhibit further growth. This leaves many tail ends to provide actin monomers but only a few leading ends to grow, thus providing a larger concentration of raw materials.

TRAFFIC ACROSS MEMBRANES

Cells require an infrastructure for containment, because biological nanosystems are typically composed of many individual freely interacting parts. If we choose to design nanomachinery with the same paradigm, we will also need effective containers. Membranes are the primary structures used for containment in natural biological systems. They have attractive properties: Lipid membranes are flexible, self-healing, and impermeable to the molecules that must be contained. However, use of membrane-enclosed spaces creates a new problem: the need to transport objects across the barrier. A perfectly sealed membrane is useless in most applications. In answer to this challenge, cells build a wide variety of active and passive transport systems to traffic molecules across membranes.

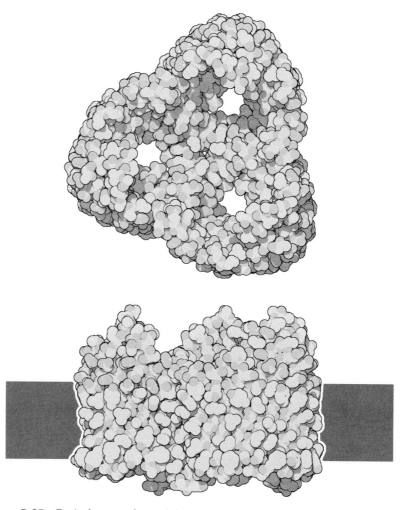

Figure 5-35   Porin forms a channel through the outer membrane of bacteria, allowing free passage of small molecules such as amino acids and sugars. It is composed of three identical subunits, each with its own membrane-spanning channel.

Potassium Channels Use a Selectivity Filter

Channels are passive transport devices, allowing free flow of molecules through membranes. In the simplest case, a channel is simply a protein with a hole through it, linking the two sides of a membrane. The bacterial protein *porin* is an example (Figure 5-35). It uses a hollow cylinder of β-sheets to create a channel, allowing the passage of small molecules such as sugars and amino acids through the membrane. As you might imagine, this is only used in the outer membrane of the bacterium, to make it porous to nutrients. The inner membrane must remain impermeable, because it is used for energy production with electrochemical gradients.

However, the most useful channels are designed to allow passage of a specific molecule. In many cases, they are also *gated*, with the ability to open or close in response to some signal, such as the binding of a specific molecule or ion, a change in the electrical potential across the membrane, or mechanical stresses. This level of control is particularly useful, and over a hundred examples of natural gated channels have been described. The molecular details are only now being determined. The atomic mechanism of one example, the potassium channel, has been revealed in crystallographic structures.

Potassium channels allow passage of potassium ions but block passage of sodium ions and chloride ions, which are also common in the cellular environment. The blocking of chloride ions is not difficult because they are negatively charged and potassium ions are positively charged. By adding a few negative charges at the entry to the channel, chloride will be repelled and will not pass through the channel. But blockage of sodium ions is a far more difficult task. Both sodium and potassium ions carry a positive charge, so an approach based on charge will not work. A simple filter based on size also will not work, because sodium ions are slightly smaller than potassium ions (0.095 nm for sodium and 0.133 nm for potassium). The trick used in natural potassium channels is to take advantage of the water environment of biological systems. In solution, ions are surrounded by a strongly associated shell of water molecules. The potassium channel is designed with a pore that is small enough to pass the ion but not the shell of waters. The channel contains several rings of oxygen atoms, formed by amino acids surrounding the channel, that mimic the shell of waters. As ions enter the narrow channel, they shed their waters but enter into an environment that is just as favorable, surrounded by the channel oxygen atoms. The ion may then exit at the other side, picking up a new shell of water molecules as it leaves the channel. The process is driven by a concentration gradient.

Potassium ions flow freely through the channel at rates of up to one hundred million ions per second. But it is also remarkably selective. The selectivity is provided by the shape of the channel (Figure 5-36). The oxygen atoms are designed to fit exactly to potassium ions, forming strong interactions from all sides of the channel. Sodium ions, on the other hand, are too small to form stable interactions with all of the surrounding channel oxygen atoms. The water shell of sodium is slightly smaller than that around potassium, so if it sheds its shell it will take an energetic loss, because it cannot form interactions with all of the oxygen atoms in the channel. This differ-

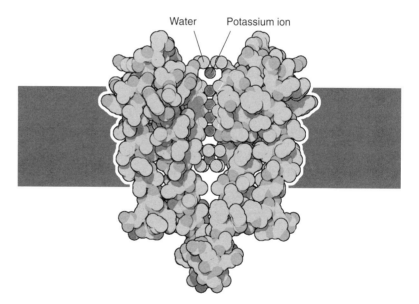

Figure 5-36 The selectivity filter of the potassium channel is formed by a ring of oxygen atoms surrounding a narrow channel. The ring is exactly the right size for a potassium ion but slightly too large for a sodium ion. In this structure, there are four potassium ions lined up in the selectivity filter and two other potassium ions, one at each end of the channel. Compare the arrangement of waters around the lowest potassium ion with the arrangement of oxygen atoms from the protein around the four potassium ions passing through the channel.

ence in energy provides the specificity, allowing only one sodium ion to pass for every ten thousand potassium ions.

ABC Transporters Use a Flip-Flop Mechanism

Transport proteins take an active approach to moving molecules across the membrane. They typically bind to a molecule and then change shape, forcing the molecule across in the process. Many mechanisms are used to power these transporters. In some cases, ATP is used to power the change in shape. Bacteriorhodopsin, described below, uses light to power transport of hydrogen ions. In other transporters, two molecules are *symported* at the same time. For instance, the lactose permease of bacteria transports the sugar lactose and a hydrogen ion at the same time, using the energy of the electrochemical gradient of hydrogen ions to concentrate lactose inside the cell. Hundreds of different transporters have been studied, providing a powerful collection of premade pumps for use in bionanotechnology applications.

The most common types of transport proteins used in modern cells are termed *ABC transporters*, where ABC refers to an ATP-binding cassette that is found in all of the examples (Figure 5-37). These proteins are shaped like a clothespin that spans the membrane. The ATP binding domains are found on the handles, and the transport mechanism includes a pocket in the jaws. The pocket has a gate on each side of the membrane. The protein is thought to undergo a "flip-flop" mechanism, where the jaws open and close. In one state, the gate is open on one side; in the other state, the opposite gate is open. The transition between the two states is powered by cleavage of ATP. ABC transporters that translocate many different molecules, such as amino acids, ions, sugars, vitamins, and toxins, have been discovered.

Bacteriorhodopsin Uses Light to Pump Protons

As discussed above, electrochemical gradients can be used to power diverse biomolecular processes. Several different methods are used to create these electrochemical gradients. The powerhouse in cells is the electron transport chain, which uses the flow of electrons stripped from sugar molecules to power the transport of electrons across the membrane (see Figure 5-10).

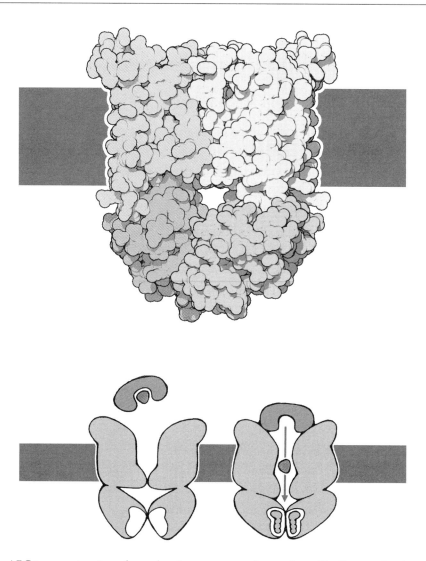

Figure 5-37 ABC transporters transfer molecules across membranes by a flip-flop mechanism. The bacterial BtuCD transporter moves vitamin B_{12} across the membrane. The transporter has a large cavity that is normally open to the outside of the cell. The vitamin molecule is delivered by a small carrier protein. Binding of ATP to subunits inside the cell causes a large change in shape, opening the base of the cavity and allowing the vitamin molecule to enter.

These proteins contain a string of electron-carrying cofactors, ranging from weak carriers to carriers with strong affinity for electrons. As the electrons travel from unstable to stable carriers, the energy of the flow is used to translocate protons across the membrane. Two mechanisms are proposed: They may guide the docking of a proton-carrying cofactor, orienting it first on one side of the membrane to pick up protons, then moving it to the other side to release them. Or the flow of electrons may force allosteric changes in the protein structure similar to the ABC transporters, opening gates on one side and then the other.

Bacteriorhodopsin is one of the best-understood proton pumps. Powered by light, the protein cycles through three states of different energy (Figure 5-38). The resting state is intermediate in energy and has a conformation that picks up a proton on one side of the membrane. It spontaneously converts to the second state at lower energy, shifting the conformation of. the molecule and moving the proton to the other side. Absorption of light converts the complex into a third form of highest energy, which forces the release of proton. The complex then spontaneously falls into the original intermediate-energy form, ready to pick up another proton. The cycle occurs only in one direction, pumping protons in one direction across the mem-

---

Lessons from Nature

- Channels through membranes may be created through proteins embedded in the membrane.
- Selective channels may be created by tailoring the chemical properties of the channel. Potassium channels use a selective stabilization after desolvation to pass potassium ions but block sodium ions.
- Specific transporters may be created with a flip-flop mechanism that successively opens gates on each side of the membrane.
- Proton pumps use a cyclic process to pick up protons on one side, transfer them, and release them on the other side. Biological examples include pumps powered by electron flow or light.

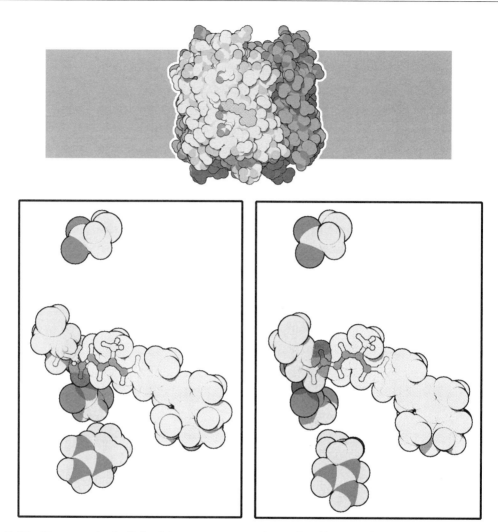

Figure 5-38 Two steps in the light-driven proton pumping cycle of bacteriorhodopsin are shown here. The structure on the left is in the ground state, before it has absorbed light. The retinal (the long molecule in pink crossing through the center) is in the *trans* form, as shown by the zigzag region shown in ball-and-stick representation. Note that the key hydrogen atom, shown in red, is pointing down. The structure on the right shows the molecule after absorbing light. Note that the retinal now has a bent *cis* shape at the site indicated by the balls and sticks. This new shape has changed the orientation of the nitrogen at the end of the retinal, pointing the hydrogen straight up. It has also shifted the position of several protein amino acids that are along the pathway of proton transfer, shown in gray.

brane, because of the need for light energy for the transition from low-energy to high-energy conformations.

The key step is a change in conformation of the small cofactor molecule *retinal*, which is induced by absorption of light. A similar change in shape is used in visual sensing (described below). Retinal contains a string of conjugated double bonds that absorb a wide range of visible light. When light is absorbed, one double bond switches from the straight *trans* form to the bent *cis* form. Retinal is connected covalently to the surrounding protein, so this large change in shape is transmitted to the protein, changing its conformation and promoting release of the proton.

BIOMOLECULAR SENSING

Our environment is full of interesting properties, such as light, sound, and pressure, that we would like to monitor, but which have very little effect on most physical objects. To sense changes in these environmental properties, we need to transduce the changes into a form that we can recognize. In biological systems, sensing is often performed by a receptor protein, which senses changes in some environmental property and then transduces this stimulus into a signal that may be recognized by the biomolecular signaling machinery. Most often, the sensory function induces a change in shape or a change in the charge distribution of the receptor protein, which is then recognized and amplified, ultimately making intracellular changes or initiating a nerve impulse.

The challenge for bionanotechnology is to harness these receptors when separated from their biological context. These receptors use sensitive recognition and amplification schemes to transduce a small molecular change in the receptor into a large cellular or nervous response. In some cases, the application to bionanotechnology will be straightforward. For instance, some receptors are channels that open on sensing. These channels may be used in any application that has two reservoirs separated by a membrane. The receptors of vision and smell, however, require a sophisticated protein-protein signaling scheme, so applications will necessarily be more complex.

Smell and Taste Detect Specific Molecules

Bionanomachinery is particularly well suited for the senses of smell and taste. Smell and taste rely on the recognition of a given molecule within a mixture of many molecules. As we have seen, this type of molecular recognition is a finely honed ability of proteins. Cells build a large collection of smell and taste receptor proteins, each of which binds to a given molecule and initiates a nervous signal when they sense their targets. The details of these useful sensors are only now being characterized.

Smell is recognized by odorant receptors. Mammals have about 1000 different genes encoding odorant receptors, which, by combination, can sense billions of different odors. The receptors are membrane-spanning proteins found on the surface of olfactory cells. Looking at the amino acid sequences of many of the genes, highly variable regions have been found in a region about a third of the way into the membrane. It is thought that amino acids facing outward from the protein form a pocket within the membrane that binds to odorant molecules. The location of the binding site within the membrane is appropriate, because smell typically senses the presence of hydrophobic, volatile organic molecules. Binding of odorants induces a change in the shape of the receptor on the inner side of the membrane. This change is the transduction event that activates the signaling chain inside the cell.

Taste includes the sensing of a variety of very different molecules—sweet, bitter, salty, sour—and uses different nanoscale approaches for each. Taste receptors for bitter substances recognize organic molecules that may be of danger to the organism, such as caffeine, nicotine, and strychnine. Many bitter receptors are receptors similar to odorant receptors. Similar receptors are also used for sweet sensing, by binding to sugars and initiating a signal. The umami taste, which gives the satisfying taste to proteins, is based on the sensing of the amino acid glutamate. It is thought to be sensed by a receptor similar to the glutamate receptor used in nerve synapses.

Salty and sour tastes are sensed differently. Because they involve sensing of the levels of ions—sodium ions for salty and hydrogen ions for sour—an ion-gated channel is used for each. These channels open when exposed to high concentrations of the ions, allowing ions to flow across the membrane. The subsequent reduction of the electrochemical potential across the membrane is the transduction event that triggers the nerve impulse.

Light is Sensed by Monitoring Light-Sensitive Motions in Retinal

Biological sensing of light relies on retinal, a small molecule that changes conformation when it absorbs a photon. Retinal contains a string of conjugated double bonds, which make it an excellent chromophore. It has a broad absorption band in the visible region of the spectrum with a peak at green light of about 500 nm. It has a high extinction coefficient, close to the theoretical maximum possible for an organic compound in the visible region. By changing the chemical environment around the retinal, the absorption spectrum can be shifted, allowing the sensing of different colors of light.

In the eye, retinal is associated with the protein opsin (shown in Figure 2-13), which is found embedded at high concentration in special membranes inside the retina. Opsin is composed of a single protein chain that forms seven α-helices. These form a cylindrical bundle that passes through the membrane. Retinal is attached in the middle of the bundle of α-helices through a linkage to a lysine amino acid. As in bacteriorhodopsin (Figure 5-38), the linkage transmits the retinal shape change to the surrounding protein.

In rhodopsin from the human eye, retinal begins in the form of 11-*cis*-retinal. When it absorbs a photon, it changes in picoseconds to the all-*trans* form. The protein then undergoes a series of changes, pushed by the new shape of the chromophore. This change in shape triggers a cascade of amplification steps that ultimately trigger a nerve signal. Finally, a change in protonation state in the linkage allows the retinal to dissociate, where it is restored to its 11-*cis* form by other enzymes and replaced into the protein.

Mechanosensory Receptors Sense Motion Across a Membrane

Study of mechanosensors has revealed some exciting concepts, but molecular details are still under study. Mechanosensors detect mechanical forces, such as touch, acceleration, and sound. In the systems under study, sensing is very fast and highly sensitive, so the receptors are thought to be ion channels. These channels open quickly and allow many ions to pass, amplifying a small signal into a significant cellular response. The basic model for a mechanosensory channel has attachments on both sides of the membrane. On the outside the channel is attached to an extracellular anchor, and on the inside it is attached to the cytoskeleton. Motion of the extracellular anchor stretches the extracellular link, causing a change in the channel and causing it to open transiently. Many of these systems show adaptation, where a constant force, such as the constant pull of gravity, is ignored and only changes

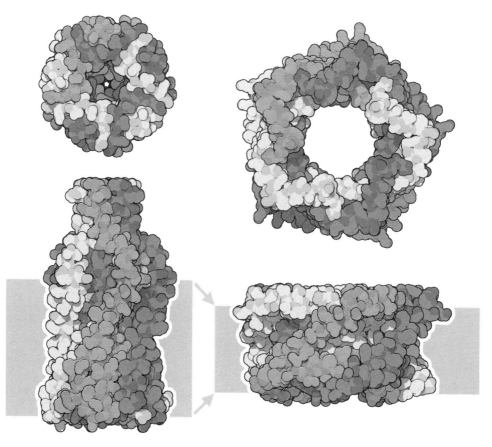

Figure 5-39 The MscL channel opens in response to stretching of a membrane. As the membrane thins, the channel opens like an iris, allowing ions to pass.

in forces, such as a small change orientation, produce signals. This is conveniently performed by relaxation of the channel and the intracellular link, progressively relaxing the stretch imposed by the extracellular link.

The best-characterized mechanosensory channel is the bacterial MscL channel, which has a simpler mechanism (Figure 5-39). This channel opens when the surrounding membrane is mechanically stretched. This is useful when bacteria are transferred to environments with low ionic strength, such as when it rains. Without the channels, osmotic pressure would build rapidly, swelling the cell and bursting it. However, the channels open as the cell swells, and ions leave the cell, balancing the ionic strength.

The MscL channel is composed of a ring of five small proteins. Each has two α-helices that cross the membrane. In the closed form, these pack tightly against each other, and a ring of hydrophobic amino acids form a tight gate, with diameter of about 2 Å, that seals the pore. When mechanical stress is applied to the membrane, the channel opens to form a pore of about 40-Å diameter, large enough for ions, metabolites, and even small proteins to cross. It is thought that the helices slide relative to each other, opening like a camera iris. This channel is highly robust. It may be purified and reconstituted in pure form in lipid micelles. It appears to act simply through sensing of mechanical stress within the lipid bilayer, without attachment to structures inside or outside of the cell. Thus it presents attractive features for applications in bionanotechnology.

Lessons from Nature

- Biosensors transduce stimuli in two ways. They may undergo a change shape in response to stimuli, which is detected by other proteins in the signaling pathway. Others are gated channels that open when they receive the stimulus.
- Sensing of specific molecules is performed by creation of specific binding sites in receptors.
- Sensing of light is performed by monitoring the state of retinal.

Bacteria Sense Chemical Gradients by Rectification of Random Motion

Bacteria are faced with an interesting dilemma. They must actively seek out food, moving from regions poor in nutrients to richer territory. But the small size of bacteria poses a problem: How do bacteria sense the gradient of nutrients, so that they can move toward regions of higher concentration? The change in nutrient concentration from the front of the cell to the back is too small to be sensed. Therefore, a temporal sensing mechanism is used. Nutrient levels are sensed at different times and then compared, watching to see whether they are increasing over time or decreasing. However, they only have information on whether the concentration is getting better or worse at any given moment—they do not save any information on the best direction to travel. So bacteria have developed a method for rectifying random motion that allows them to travel up concentration gradients in their environment.

The flagellar motor of bacteria may turn in either direction. In one direction, several flagella form a bundle that drives the bacterium in one direction. In the other direction, however, the flagella separate and the bacteria tumbles rapidly in place. By combination of these two behaviors, bacteria follow concentration gradients. If the cell senses that the nutrients are increasing in level, the cell swims steadily. If the cell senses a drop in levels, however, the flagellar motors reverse, and the cell tumbles, picking a new direction to swim. The result is a biased random walk, which eventually works its way up the gradient.

SELF-REPLICATION

Self-replicators are seen by many to be a necessary part of any nanotechnology scheme that aims to create macroscale objects. This is a consequence of the scaling of nanoscale construction. A single assembler cannot construct materials fast enough to create macroscale objects in a reasonable amount of time. Building twice as many assemblers, or a thousand times as many, does not help—the growth is only additive. There are just too many individual molecules in a gram of material. To span the gulf between the nanoscale and the macroscale of our familiar world, we must turn to self-replicators, which give exponential growth.

Self-replicating entities have great promise but also present great perils. Life on Earth provides vivid examples of both sides of this coin. The interconnected harmony of a prairie or a mountain ecosystem is a brilliant examples of the beauty that is possible, even with self-replicating entities that are each struggling and competing according to their own goals. The spread of AIDS and the blight of kudzu are examples of the dangers of rogue self-replicators. Nature gives us the tools to create custom self-replicating bio-nanomachinery, but this knowledge must be continually tempered with caution.

Cells Are Autonomous Self-Replicators

Self replication has proven to be an amazingly successful approach, as shown by the ubiquity and diversity of life on Earth. All cells on Earth are built according to a similar molecular plan, using a similar mechanism to self-replicate. A very simplified parts list includes the following.

(1) *Information-driven assembler.* This nanomachine builds new molecules according to the information held in some form of storage medium. The ribosome is the molecular assembler used by modern cells.

(2) *Information storage medium.* This medium includes instructions for the assembler, directing the assembly of all of the other parts. DNA is the information storage material used in modern cells.

(3) *Duplicator.* This nanomachine duplicates the information storage medium. DNA polymerase performs this task in modern cells.

(4) *Chemical processors.* These nanomachines convert available raw materials into the building blocks used by the assembler and the duplicator. Modern cells require thousands of enzymes to perform these chemical transformations.

(5) *Infrastructure.* These nanomachines provide the infrastructure to define the body of the cell, separating it but still allowing communication with the environment. This infrastructure also includes the nec-

essary machinery for reproduction by binary fission. The basic infrastructure of modern cells is a membrane composed of lipids and its associated transport and signaling proteins.

Organisms are remarkable: Their seemingly boundless complexity is defined by a small instruction set. An entire human being is specified by the information in a remarkably small genome and the structure of a single cell. Remarkably, we are the first generation able to read this amazing text, and to harness this information for our own use. Today, we have access to our own blueprint and the blueprints for bacteria, viruses, flies, and plants. We can now work to understand how the thousands of proteins work together within our cells, and how we might modify them to perform new industrial and medical tasks.

Current estimations of the minimal number of genes needed to create a living organism that will grow and reproduce under laboratory conditions are in the range of 250 to 350. The simplest organisms, the mycoplasmas, have about 550 genes (Table 5-1). However, these organisms live inside other cells and rely on their hosts for many of the basic biosynthetic precursors. They retain basic functions such as protein synthesis and management of DNA, energy management, and a core set of cell-homeostasis proteins. They also contain a significant number of transporters for

Table 5-1   Parts List for *Mycoplasma genitalium*

Function	Number of Genes
Synthesis of building blocks	37
Housekeeping	21
Energy	31
Regulation	7
Protein synthesis	145
Transport and delivery	34
Cell Envelope	17
Others	27
Unknown function	152

Source: Adapted from Fraser, C.M. et al. (1995) "The minimal gene complement of *Mycoplasma genitalium*." Science 270, 397–403.

handling the movement of molecules across the cell membrane. Comparison with other organisms and experiments that selectively destroy the function of proteins have shown that only about one-half of these are absolutely necessary.

Of course, the genome is not the only information that specifies cellular structure and function. The genome must be placed within a living cell to be effective (Table 5-2). In the simplest cells, a single membrane enclosing the cellular machinery appears to be enough. This provides two functional compartments: the interior, soluble space and the membrane with its embedded proteins. In our cells, however, a complex hierarchy of organelles must be passed from generation to generation to provide the seed for growth. There is no indication, for instance, that our cells could build an entire mitochondrion from scratch. So the blueprint for building a cell must include the information held in the genome, along with a map of the structure of a living cell. This provides enough information to specify the genetic information and the mechanism needed to use it for self-replication.

Table 5-2 Composition of a Typical *Escherichia coli* Cell

	Number of Molecules	Number of Different Types
Protein	1,330,000	1,000
Ribosomes	18,700	1
Transfer RNA	205,000	60
Messenger RNA	1,380	400
DNA	2	1
Lipid	22,000,000	4
Lipopolysaccharide	1,200,000	1
Peptidoglycan	1	1
Glycogen	4,360	1

	Percentage of Weight
Macromolecules	28.83
Small molecules	0.87
Inorganic ions	0.3
Water	70.00

Source: Adapted from Neidhardt, F.C., Ingraham, J.L., and Schaechter, M. (1990) *Physiology of the Bacterial Cell: A Molecular Approach*, Sinauer Associates, Sunderland, MA.

Lessons from Nature

- Self-replicating cells contain five basic functionalities: an information-driven assembler, an information storage medium, an information duplicator, a set of synthetic machinery for creating construction materials, and a general infrastructure.
- Using the design of modern cells, about 250–350 genes are needed for self replication.
- The design of modern cells is limited by the process of evolution. Other designs for self-replicating entities may prove more efficient.

The Basic Design of Cells Is Shaped by the Processes of Evolution

Life on Earth uses a basic template for creating self-replicating entities: a combination of information-driven assembly, synthetic machinery for building blocks, and containment infrastructure. This design for building a self-replicating entity is strongly molded by the evolutionary process under which life developed. To date, we have stayed close to this design, using and modifying existing natural systems to perform our new applications. But in the future, we may choose to take different approaches.

In particular, the need to enclose an organism and separate it from the environment is a product of evolutionary selection. It is necessary to enclose genetic information within a defined cell so that the cell may compete with other cells. If, for instance, different assemblers were floating free, the best assemblers would be building copies of themselves but also copies of competing plans. Instead, it is necessary that the best assemblers only work toward reproducing themselves.

However, if the need for evolutionary selection is removed, the requirement for containment may also become unnecessary. If we are doing the designing, we don't have to worry about competition and we can simply build the machine of the best design. One might imagine a "gray goo" composed not of self-contained nanorobots, but instead of a collection of bionanomachines that come together and perform specific tasks and then disassemble when finished.

MACHINE-PHASE BIONANOTECHNOLOGY

One of the novel concepts proposed for molecular nanotechnology is the definition of a new phase of matter—*machine-phase matter*—which is different from solids, liquids and gases. Machine-phase matter is composed of a collection of nanoscale machines with articulated components that perform ordered information-driven functions. Most natural biomolecular machinery adopts a different paradigm, more akin to chemistry than this atomically defined machine phase. Natural bionanomachines are created at the nanoscale and then combined through random diffusion inside cells to perform the more orchestrated tasks of life. There are many examples, however, in which machine-level control extends beyond the nanoscale of individual bionanomachines and extends into realms closer to machine-phase matter.

Natural machine-phase assemblies are highly robust. They are composed of many modular nanomachines, each highly stable and functionally efficient. These assemblies are also highly redundant, employing many copies of each bionanomachine. Often, two levels of hierarchy are used: first assembly of nanomachines into cellular assemblies and then assembly of cells into macroscale devices. At both levels, a high degree of redundancy is used when they are combined into the functional complex, providing benefits in production and error control.

In the following section, I present two familiar examples of machine-phase constructions used by modern organisms. Parts of each of these have been used in bionanotechnology. But, unfortunately, we are not yet able to perform the feat accomplished by these cells, constructing entire nanoassemblies of this complexity and precision.

Muscle Sarcomeres

One of the most spectacular examples of machine-phase matter in natural systems is the muscle sarcomere (Figure 5-40). Muscles combine the action

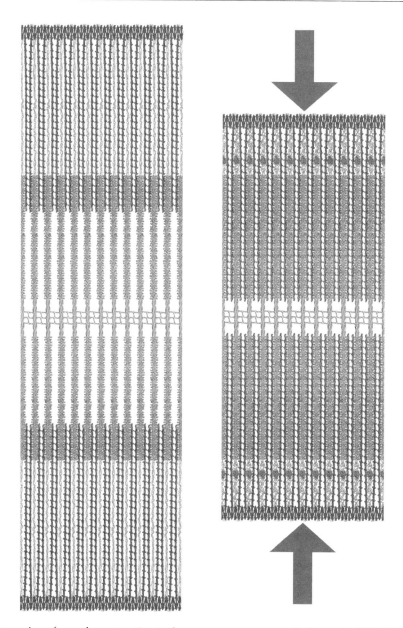

Figure 5-40 The engine of muscle contraction is the sarcomere, composed of myosin thick filaments, shown here in pink, and actin thin filaments, shown in gray. The actin filaments are held in place by a dense network of proteins collectively known as the Z-line, shown at the very top and bottom, which connects this sarcomere to others above and below. The myosin filaments are bipolar, with myosin facing in opposite directions at the two ends. The myosin head groups climb along the actin filaments, contracting the entire sarcomere. The thin, snaky molecule connecting the myosin filaments to the Z-line is titin.

of many myosin molecules to create macroscopic motion. "Many" is an understatement—your biceps contain some 5×10^{19} myosin molecules, perfectly arranged to work in concert to move your arm on demand. The structural unit creating contractile force is termed a *sarcomere*, which is composed of thick filaments composed of myosin interdigitated between thin filaments composed of actin. Each sarcomere is about 2–3 μm long and 1–2 μm in diameter, and many sarcomeres are attached back-to-back to fill muscle cells.

The thick filament is composed of several hundred myosin molecules, each of which has two motor domains. Unlike kinesin, only one is active at any given time. They reach from the thick filament over to the neighboring thin filament, attach and pull, and then release. The thick filament is bipolar, so force is generated in opposite directions at the two ends. The result is the pulling of actin filaments on either side toward the center. The entire structure is held together with a large collection of bundling proteins, which link together the thick filaments at their center (forming the dense "M" line seen in electron micrographs) and the actin filaments at the ends of the sarcomere (forming the "Z" line). Sarcomeres are inherently unstable structures—any small deviation of the thick filaments from the center will create a greater overlap with the actin filaments and thus a greater force on that side. The huge elastic protein titin is used to tether the thick filaments in the center of the sarcomere, solving this potential problem.

The geometry of the muscle sarcomere gives it an effective range of lengths from about 2.25 μm in rigor to 3.65 μm when relaxed, contracting by about 60% in length. Force generation is nearly linear through this range of lengths. Each crossbridge can apply approximately 0.8 pN of force, so we are using the force of about 2 trillion myosin motors when we hold a baseball in our hand.

Regulation of this contraction is controlled by proteins, termed tropomyosins, that are bound on the surface of the actin filaments. In a process controlled by calcium levels, tropomyosin shifts position, exposing the site of myosin binding. The process begins slowly with just a few myosin heads going through their mobility cycle slowly, but as more myosin binds, tropomyosin is shifted to its fully active position and the actin filament becomes fully active as a track.

To form muscles, sarcomeres are stacked end to end, forming a ma-chine-phase nanomachine that is centimeters in length, and working in se-ries to provide centimeter-length contractions. In skeletal muscles, many cells fuse to form long multinucleated muscle fibers. These are packed full of sarcomeres, surrounded by the machinery of ATP production and calci-um regulation.

Nerves

In natural systems, programmable biological computing occurs at the level of cells, not at the nanoscale level of molecules. Genetic information is es-sentially hard-wired for a given organism, and all operations performed with genetic data are essentially fixed, except on evolutionary time scales. As described in Chapter 6, nanoscale programmable computing is currently possible and is being developed with many approaches. But nature does its programmable computing at the microscale with neurons.

Neurons are programmable electrical components allowing all of the fa-miliar programmable capabilities of computers. The well-studied worm *C. elegans* has a mere 308 neurons, whereas the human brain contains up-ward of 100,000,000,000 neurons. Each neuron is composed of an input lay-er, the cell body and associated dendrites, and an output layer, the terminal branches, connected by a line of high-speed electrical communication, the axon. Individual terminals at the output layer communicate with individual sections of the input layer of other neurons across "synapses," sending ei-ther an excitatory or an inhibitory signal. The summation of all of the inputs to an individual neuron determines whether it sends a signal down its own axon. The programmable aspect of neurons is thought to occur at the level of communication from the terminal to the input cell, where synapses that show high, rapid usage rates are modified to send stronger signals at later times. Typical human neurons may have a thousand output terminals and may take up to 100,000 inputs from other neurons, providing an unparal-leled level of computational complexity. Remarkably, researchers are now using discoveries from nerve development to grow neurons into custom configurations, opening the possibility of creating custom neural networks.

Looking at the nanoscale level, the machinery used for nerve signal transmission provides many opportunities for application to bionanotechnology. It includes machine-phase mechanisms for signal transmission using electrochemical gradients and small-molecule messengers. It employs an effective lipid insulator. It uses an efficient analog-to-digital converter, described in more detail below. These molecular machines are quite robust. For instance, researchers have isolated a large axon from squid and squeezed out all of the cytoplasm inside. The remaining membrane can be refilled with salt solutions and made to carry artificial nerve signals. This machinery may be harnessed in whole or in parts for bionanotechnology.

An example of analog-to-digital conversion is particularly fascinating. The action potentials that pass down nerve axons are intrinsically digital. When resting, axons have a electrochemical gradient across the membrane, with an excess of sodium outside. The message is passed down the membrane by using this potential for power. The membrane contains many voltage-gated sodium channels. These channels are normally closed, but when the potential of the membrane is reduced (depolarized) they open, allowing sodium to cross. The result is a wave of depolarization down the axon. A few channels open at one end, which depolarizes the area, opening channels next to it, and so on. The resultant action potential passes down the axon. However, because the signal opens all of channels along the path of the axon, it cannot carry any information about the magnitude of the original event that started the signal. But the magnitude of this signal is important, because it is a reflection of the sum of the many signals that are being received from all of its inputs. So how does the neuron convert an analog signal into a digital signal?

Lessons from Nature

- Natural biomolecular systems use a diffusion-to-capture paradigm for most functional tasks. This method is sufficiently fast and accurate for most nanoscale processes. Machine-phase assemblies are used for specialized microscale and larger tasks.

Neurons encode the magnitude of the signal in the frequency of action potentials sent down the axon. Two potassium channels at the starting point of the axon have the job of converting an analog signal—the summation of excitatory and inhibitor signals received by the cell body—into this frequency-encoded digital signal. When potassium channels open, they allow potassium ions to exit the cell, restoring the electrochemical gradient. One type of channel, the delayed potassium channels, have slower kinetics than the sodium channels and open after the action potential has started. They serve to restore the gradient quickly, allowing action potentials to be sent in rapid succession. A second channel, the early potassium channels, are sensitive to the level of the depolarization signal. If it is weak, they open, weakening the electrochemical gradient and increasing the time before it is possible to send another action potential. Thus if the starting signal is weak, action potentials are spaced further apart in time.

BIONANOTECHNOLOGY TODAY

6

*Now, you might say, "Who should do this and why should
they do it?" Well, I pointed out a few of the economic
applications, but I know that the reason that you would do
it might be just for fun.*

—Richard Feynman

Bionanotechnology is a reality today; in fact, it is a booming field. It's an exciting time to be working in bionanotechnology. Everything is new, and new discoveries are reported every day. It is a particularly exciting time to be working in this field because bionanotechnology today is driven by clever people. A clever new idea can open an entirely unforeseen avenue of research and application. The first glimmerings of nanomedicine are allowing researchers to make tailored changes to the mechanisms of the human body, correcting defects and curing disease. Familiar biomaterials, such as wood, bone, and shells, are providing the principles needed to create materials tailored at the nanoscale to fit our needs. Biological methods of nanoscale information storage and retrieval are being harnessed to solve computational problems and convict criminals. In this chapter, we will look at a few examples of current bionanotechnologies.

Bionanotechnology: Lessons from Nature. By David S. Goodsell
ISBN 0-471-41719-X   Copyright © 2004 John Wiley & Sons, Inc.

BASIC CAPABILITIES

Biology provides a plethora of basic methods for the construction of functional nanomachinery. Researchers have built upon these natural raw materials and are now exploring a wider set of construction strategies. These will be important as we move further away from biology-inspired nanomachinery and begin to create nano-worlds of our own.

Natural Proteins May Be Simplified

Natural proteins are daunting. As researchers are discovering, there is no simple relationship between the sequence of amino acids in a protein chain and the final folded structure. Our preferred approach to engineering is more parsimonious; we like to design machines with the most efficient combination of parts. Our macroscale machinery is built of tidy plates and gears and axles, all held together with screws that fit into precisely machined holes. Quite naturally, bioengineers have attempted to simplify the design of proteins, to allow the level of control that we enjoy with macroscopic engineering. Quite surprisingly, this has been successful. In many cases, the structures of proteins are quite robust even in the face of major design changes.

Some of the earliest experiments were performed on the enzyme ribonuclease by Bernd Gutte. He set out to find a minimal chain that would still fold and perform the cleavage reaction. He began with the structure of natural ribonuclease, which contains 124 amino acids, and noticed that the active site is composed of only a small core structure of the protein. Many of the loops extending from this core seemed to be unnecessary. So he saved the core structure and then clipped off all of the large loops that extend from the core, replacing them with smaller loops. He was able to design an enzyme that included only 63 amino acids but still folded and performed the catalytic reactions (Figure 6-1).

Natural proteins are built from 20 different amino acids. With the exception of key reactive amino acids in enzyme active sites, it is often not apparent why certain amino acids are chosen in different places in protein structures. It would be much simpler to pick a reduced set of perhaps five or six amino acids to create the basic infrastructure of a protein and save the

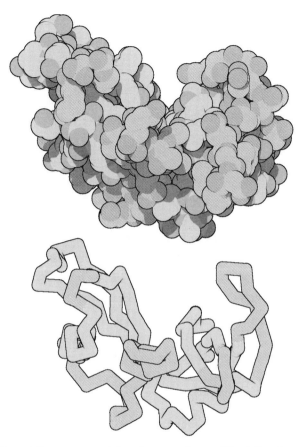

Figure 6-1 Bernd Gutte removed four loops and one of the tails, shown in pink, of ribonuclease, replacing them with a single alanine or glycine amino acid. The result was a half-size enzyme that still performed its RNA-cleaving function.

most exotic amino acids for their specialized functions. David Baker and coworkers have tested whether this approach is feasible. They started with a small protein domain that is widely used in signaling proteins. They attempted to design this domain with a limited set of amino acids. Again, they were surprisingly successful. Using site-directed mutagenesis and methods for screening thousands of different changes, they replaced each amino acid in the protein by isoleucine, lysine, glutamate, alanine, or glycine. After testing many changes, they found a simplified version of the protein that folded into a stable structure and retained much of its function.

In this simplified protein, 40 of the 45 amino acids in the chain were changed to one of this set of amino acids.

Proteins Are Being Designed from Scratch

Of course, to take full control of bionanotechnology, we must be able to design a protein. We must be able to predict the proper amino acid sequence that will fold into any desired structure. Currently, we can predictably make small local changes to known protein structures with homology modeling, to tailor the local characteristics of a protein. However, to allow the necessary freedom to explore entirely new applications, we need to have the ability to design proteins using only basic principles.

Since the early 1980s, several pioneering laboratories have been designing proteins from scratch. As more and more information has been gathered on the forces and geometries that stabilize protein structure, successes in *de novo* protein design have multiplied. The most common current approach is to start with a desired protein fold, such as a bundle of α-helices, and then to design an amino acid sequence that will stabilize the folded structure. Researchers have discovered that negative design is critical to this process. The amino acid sequences must be designed to avoid improper competing conformations.

Jane and David Richardson were among the first to attempt this challenge. They designed two proteins based on the basic folding patterns of natural proteins. But instead of using homology modeling with the existing protein structures, they attempted to design a novel amino acid sequence with no homology to any existing proteins. Their first protein, "Felix," was modeled after proteins composed of bundles of four α-helices. It contains 79 amino acids and includes 19 of the 20 amino acid types. When synthesized, it was found to be soluble and showed a high content of α-helix when analyzed by circular dichroism spectroscopy. However, they found that it is only marginally stable, and when analyzed by NMR, it shows a significant amount of disorder. Their second protein, "betabellin," was modeled after proteins with two sandwiched β-sheets. After a dozen redesigns providing valuable lessons at each step, a stable, soluble version has been obtained.

William F. DeGrado and coworkers have had great success in designing proteins composed of bundles of α-helices, designing examples that show a stable, folded structure reminiscent of natural proteins. One of their design efforts began with the structure of a simple peptide that crystallizes in a three-helix bundle. They then designed a protein based on the alignment of these three short α-helices, adding loops to connect the three peptides into one continuous chain. The design effort proceeded through several modifications. The helices needed to be shortened to the size found in typical globular proteins. The amino acids at the ends of the helix had to be chosen to form a suitable cap. In addition, the core formed between the three α-helices was carefully redesigned to incorporate specific charge pairs and proper packing of carbon-rich amino acids. The result was a protein, termed "α3D," composed of 73 diverse amino acids that was stable enough to yield an NMR structure (Figure 6-2).

Bassil Dahiyat and Stephen Mayo have automated this process, using the computer to design proteins that fold into stable structures. They start with a protein folding pattern, in this case, a combination of a small β-sheet

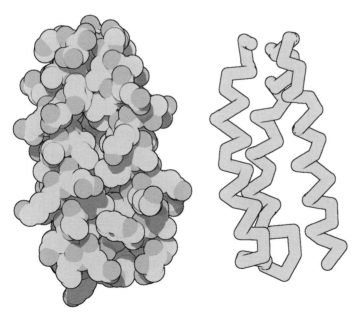

Figure 6-2 α3D is a small protein designed from scratch. It is composed of three α-helices connected by short loops.

and an α-helix that is seen in zinc finger motifs that bind to DNA. They then use a searching algorithm that tests all possible combinations of amino acids, scoring each combination with a function that evaluates the different contributions of amino acids at the surface and buried inside. After screening a library of 1.9×10^{27} possible combinations, a protein termed "FSD-1" was obtained. When synthesized, it folded into a compact structure just as predicted by the computational algorithm.

Of course, once we can design a stable folded protein, we would like to have it do something. In recent work, DeGrado and his coworkers are attempting to design a metalloenzyme based on their successful protein folds. They are approaching this in two steps. First, a stable protein structure was designed that includes a metal-binding site, using an approach similar to their three-α-helix protein described above. They achieved this goal with the protein "DF1," which contains a collection of glutamate and histidine amino acids that coordinate two metal ions. Unfortunately, it was found that the site was buried in the protein. To incorporate the metals, the protein needed to be denatured and then refolded in the presence of metal ions. In the second step, they reduced the size of amino acids that were blocking the metal-binding site. This change produced an active site that allows access to the metal-binding site and also forms a pocket for binding of substrates. The new protein, termed "DF2," allows free access to the metal-binding site in the folded protein.

Proteins May Be Constructed with Nonnatural Amino Acids

Natural proteins are limited to the 20 natural amino acids, occasionally augmented with a few modifications performed after the protein is made. Although this provides a remarkable breadth of function, as presented in Chapter 5, some applications need additional chemical or structural diversity. Natural proteins typically use prosthetic groups to perform these extended functions. However, prosthetic groups can be complicated to synthesize and to incorporate into proteins. So, quite naturally, researchers have developed methods to extend the range of amino acids that are incorporated into proteins.

The first methods for incorporating nonnatural amino acids at specific sites in proteins were developed in the early 1980s and have been greatly expanded since then. Several methods have been employed. The tightest control, of course, is available through total chemical synthesis of the protein. Then any amino acid may be added at any position, limited only by the imagination of the researcher. However, total chemical synthesis is expensive in time and materials. For larger proteins, the protein can be constructed in small pieces, 30 to 50 amino acids in length, and then assembled into the final protein. Alternatively, if only a few nonnatural amino acids are needed, a method that combines chemistry and biology may be used. Small peptides with the nonnatural amino acid may be synthesized and then attached to the remainder of the biologically produced protein.

Biological production of proteins, however, is much cheaper than chemical synthesis, so many researchers have developed ways to incorporate modified amino acids into biologically produced proteins. In many cases, specific amino acids may be modified after the protein is constructed. One of the most common approaches is to link novel chemical groups to cysteine amino acids, using a disulfide linkage. This has been widely used in current bionanotechnology, for instance, in the artificial constructions using ATP synthase motors described below. The protein is engineered to remove any natural cysteine amino acids from the surface, and then one or two new cysteines are engineered onto the surface in appropriate places.

Surprisingly, it is also possible to trick the ribosome into adding a variety of nonnatural amino acids as it builds proteins. The trick is to create a new transfer RNA that encodes the nonnatural amino acid and use a "Stop" codon to specify its location in the protein. Cells use three different stop codons: UAG, UAA, and UGA. One of these is chosen to encode the new amino acid, and a transfer RNA is created with the matching anticodon at one end and the nonnatural amino acid chemically attached at the other end. An RNA message is then created with this codon at each site where we want to add the nonnatural amino acid. Proteins are then built with a cell-free translation system, with the new transfer RNA added into the mixture. As the ribosomes build the protein, they add the new amino acid whenever the coopted stop codon is encountered.

One problem is encountered. The cell-free translation system, which is typically an extract of the cytoplasm from a cell, also includes release factors that cause the ribosome to stop at stop codons. These will compete with the new tRNA, often causing the formation of a shortened protein chain, terminated at the sites where we want to add the new amino acid. One solution to this problem is to choose a species of bacteria that makes several different release factors. When the cell extract is made, we can destroy one of the release factors, leaving the remaining ones to perform the proper termination at the remaining two stop codons.

Despite the fact that these methods are time consuming, requiring difficult syntheses of novel transfer RNA molecules and yielding only small amounts of proteins, they have been used in many research applications. After testing many variants, the range of amino acids that are tolerated by the ribosome have been found. The ribosome will add a wide range of amino acids, some significantly larger than the natural amino acids. However, the configuration of the backbone is limited to the natural handedness and the backbone must have a single carbon atom between each peptide unit. The method allows engineering of proteins at a very fine scale, making changes of a single atom. In one study, a number of different amino acids were added at a specific location in luciferase, the protein that creates light in fireflies. These modifications subtly changed the chemical environment around the active site, changing the color of the light produced.

Peter Schultz and his coworkers have taken this concept in an exciting direction and have developed a general scheme for adding nonnatural amino acids to proteins inside living cells. Using a remarkable combination of biochemical and genetic techniques, they have engineered a living bacterium that can incorporate O-methyl-L-tyrosine into proteins. They added both an engineered transfer RNA, such as those described above, and the enzyme needed to charge the RNA with the new amino acid. They find that the new amino acid is added with remarkable fidelity. To expand the concept, they are also designing transfer RNA molecules that read larger codons of four or five amino acids. Surprisingly, the ribosome will use these artificial transfer RNA molecules, aligning them with the proper four or five bases in the message. This significantly expands the number of possible codons that may be specified for use with new amino acids. Instead of coopting one of the normal codons, a whole new genetic code of longer codons could be defined. In the search for expanded genetic codes, they are even testing changes to the DNA itself. They have found that the base 7-aza-indole will form pairs with itself in DNA helices. Remarkably, this modified base is also added correctly when DNA polymerase replicates a strand. These are the first steps toward making direct changes in the basic processes in life, creating entirely new organisms different from anything found in the natural world.

Peptide Nucleic Acids Provide a Stable Alternative to DNA and RNA

Short oligonucleotides offer many exciting possibilities in nanomedicine. When added to cells, short stretches of RNA can associate with natural messenger RNA, blocking its use by ribosomes. If the sequence of the RNA is chosen correctly, it can silence disease-causing genes, blocking the production of unwanted proteins. Small nucleic acids are also useful for diagnosis of disease: They can be used to identify compromised genes in living cells. Unfortunately, cells protect themselves with aggressive methods for destroying foreign nucleic acids, because foreign RNA molecules are often a sign of pathogenic infection or cell damage. Peptide nucleic acids, developed by the Danish group of Buchardt, Nielsen, Egholm, and Berg, offer a resilient alternative to RNA and DNA for these applications.

Peptide nucleic acids (PNA) contain the normal complement of DNA bases but use a peptide linkage instead of the sugar-phosphate backbone to connect them together (Figure 6-3). They are more stable than DNA, resisting both acidic and basic conditions that destroy DNA. Because the backbone is different from that of the natural nucleic acids, they are also resistant to the enzymes that destroy nucleic acids inside cells. Serendipitously, this chemical structure closely matches the natural structure of DNA. PNA strands form hybrid double helices with DNA strands, forming the proper base pairs and actually binding more tightly than normal DNA. Sequences must be carefully designed, however, as they may show many unwanted binding modes: They bind in parallel and antiparallel orientations with DNA and can form a range of unusual double and triple helical complexes. Although they are rather inefficient at displacing the normal double helical structure of DNA in cells, they are able to slip in during the normal un-

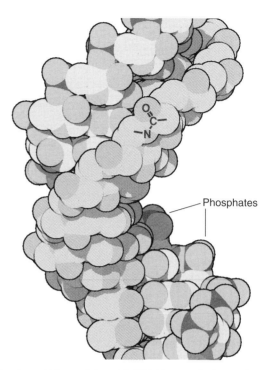

Figure 6-3 A short peptide nucleic acid (PNA) strand, shown in pink, is bound to an RNA strand, shown in gray. It forms a remarkably regular double helix, despite the large chemical difference between the peptide group in the PNA strand and the phosphates in the RNA strand. (Note: this diagram does not show any hydrogen atoms, including the one that is bonded to the amide nitrogen indicated by the chemical diagram.)

winding and exposure of bases that occurs when the DNA is read during protein synthesis. They have been shown to be potent inhibitors of transcription in cells, while being much more stable than short DNA or RNA strands.

Similar modifications have been proposed for proteins for many years. This is seen by many to be an essential step in the development of nanotechnology, to allow the construction of protein-inspired structures that are more stable than natural proteins. This is a far more difficult prospect, however, than the modification of nucleic acids. The structure of nucleic acids is driven in large part by the stacking and pairing of the bases—the backbone is, to large extent, merely a spacer that lines up the bases. Proteins, on the other hand, rely on the properties of the peptide backbone for stability. The structures of the α-helix and the β-sheet are direct consequences of the properties of the peptide unit. Changes in the amino acid module used to build proteins would invariably lead to an entirely new set of rules for folding and stabilization of the chain. Although this will not be an insurmountable hurdle as our predictive and synthetic abilities improve, it seems ambitious at this point when our ability to predict natural protein structure is currently undergoing significant growing pains.

NANOMEDICINE TODAY

One of the great promises of nanotechnology is increased control over our personal health. As our understanding of the human body has deepened, so has our understanding of disease, opening the door to therapies for treating disease. The art of surgery builds on centuries of knowledge in anatomy. The benefits of hygiene and antiseptics were made clear with the discovery of microorganisms. Today, with our growing knowledge of the atomic-scale structure of our bodies, we can now exert control at the nanoscale level.

Nanomedicine is a natural application for bionanotechnology. After all, the human body is designed for maximal function of biological molecules. This is ideal for nanomedicine, because we can use the raw materials that nature has given to us. We have incredible disease-fighting systems to use as examples. The immune system gives us tools for seeking out pathogens and quickly dispatching them. The blood clotting system gives us the tools to patch major damage in a matter of seconds, and the processes of wound healing show us how to forge lasting repairs. Now, we have the ability to tailor these tools to perform functions that nature has overlooked.

Thus far, a targeted approach to nanomedicine has been the most successful. A single target is chosen that causes (or contributes to) the disease state. A specific nanoscale device is then created to find that target and correct its function. A familiar example of this approach is aspirin. When we take aspirin, we are flooding our body with nanomachines, each composed of only 20 atoms. This nanomachine contains a recognition element that seeks out an overactive pain signaling protein and a warhead that attaches to the protein and temporarily stops its action. With aspirin, we are controlling our own bodies at the nanoscale. Currently, drug therapy is our most effective method for targeting a pathogenic organism or a given cell type or disease state.

The pharmaceutical industry is dedicated to discovering effective methods of making these molecule-level adjustments. Most common drugs were initially discovered by screening natural products. Plants and microorganisms create a remarkably useful array of small molecules, originally created as poisons and toxins for their own protection but then harnessed by us for use in medicine. Of course, once an effective drug is discovered, the mechanisms of action may be studied and improved. Today, nanoscale design adds a new level of control to the discovery and optimization of drugs. Rational design, based on our nanoscale knowledge of the target, is used to enhance existing drugs and to design entirely new medicinal compounds.

Drug design is just the first step in our long climb toward nanomedicine. Drugs have serious limitations. Because they are small molecules, they tend to target many molecules in the body instead of just the desired one, leading to side effects. As nanomachines, they have only the minimum range of functions. Today, researchers are working to create nanomachines that are more effective by incorporating methods of specific targeting to reduce side effects.

Current approaches to nanomedicine are also beginning to explore corrective therapies that can cure disease at its inception. The dream of nanorobots that sweep through the body fighting disease and making repairs is not yet a reality. However, methods are being tested for correcting specific genetic problems at their source. The ethical questions for this work are significant, but the potential rewards are too great to be ignored.

Computer-Aided Drug Design Has Produced Effective Anti-AIDS Drugs

Rational drug design is a major triumph of current nanomedicine. A target is chosen in the pathogenic organism, and the target is characterized at the atomic level. Then, using this nanoscale information, a team of scientists engineers a molecule that specifically attacks the target, blocking its action.

The team typically includes a biologist, who performs the characterization of the target and tests trial drugs; a chemist, who synthesizes drugs; and a computational chemist, who designs and optimizes drugs to bind to the target. Through successive cycles of design, synthesis, and testing, drugs are discovered, perfected, and then used in therapy.

HAART (highly active antiretroviral therapy) is a successful example of rational drug design. HIV is arguably the best-characterized organism known to science, and this knowledge has been aggressively pressed to service in the fight against AIDS. In a matter of a decade, AIDS has changed from a uniformly deadly disease to a manageable disease in many cases, because of the nanoscale design of effective anti-HIV drugs.

Several of the enzymes involved in the life cycle of HIV have been characterized and used as targets for drug design. The reverse transcriptase, which copies the viral genome into a form that is read by the infected cell, was the first to be characterized and the first that was subjected to drug therapy. Because it acts on nucleic acids, early drug design efforts focused on modified nucleotides, which are added to a growing nucleic acid strand but which have modifications that prevent further growth. Thus they prematurely terminate the copying of the viral genome, creating defective copies. Drugs such as AZT and DDI fall into this category.

With the crystallographic structure of HIV protease, true nanoscale design began (Figure 6-4). Specific inhibitors based on the natural function of the enzyme were designed. It normally cleaves the viral proteins, clipping them into the proper sizes needed for viral reproduction. So inhibitors were designed to mimic this reaction by creating a short peptide with an uncleavable bond at the site that is normally broken. The side chains on this peptide were then designed to provide maximal binding strength to the protease active site. The drug molecule mimics a peptide, but once it binds it sticks tight and blocks the action of the enzyme. Rational drug design efforts by a number of pharmaceutical companies have yielded several highly effective inhibitors of the protease. When used in combination with inhibitors of reverse transcriptase, they form the powerful triple cocktail of HAART therapy.

Treatment of HIV infection with drugs, however, quickly encounters a major obstacle. HIV mutates rapidly and quickly develops drug-resistant

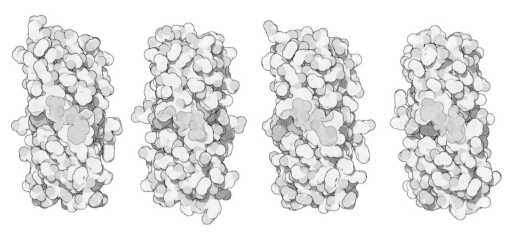

Figure 6-4 Many powerful AIDS drugs have been designed to bind in the active site of HIV protease. Four widely used drugs are shown here, from left to right: indinavir, saquinavir, ritonavir, and nelfinavir.

mutants that evade therapy. The trick is that the virus may mutate its enzymes in subtle ways that reduce the binding of drugs but retain enough activity in their natural reactions to allow growth of virus. Nanomedicine is now tackling this problem by closely examining the mechanisms by which the enzymes perform their functions and designing drugs that match these so closely that there is no room for resistance mutation. This is a challenging problem for rational drug design. In traditional drug design efforts, the researcher looks for a drug to block a single target. In the search for robust inhibitors that evade drug resistance, however, the researcher must ensure that the drug is effective against the target and all of its possible mutants.

Immunotoxins Are Targeted Cell Killers

In many therapies, we want to kill entire cells. For instance, cancer is caused by cells that are growing without control, so in cancer therapy we want to find these cells and kill them while sparing the surrounding normal tissue. What we need is a nanoscale scalpel to seek out cancer cells and remove them.

Many natural toxins, made by plants and bacteria, seem like good candidates for this function. They often kill only specific cells. This is important, after all, so that the organism making the toxin does not kill itself in the process. Many of these toxins are built from two components. One component binds to target cells, and the other component is the poison that kills the cell. Unfortunately, these toxins still attack many types of cells in an organism and would kill cells throughout the body if used in therapy. In nanomedicine, we require an even more specific approach, designing toxins that target a single cell type, such as the cells in a tumor.

Researchers have taken the approach used by natural toxins and customized it, using the fine-edged specificity found in antibodies. They have designed a hybrid molecule that links a specific antibody, such as a tumor-targeting antibody, to a cell-killing toxin. The result is an *immunotoxin* (see Figure 3-6) that will seek out cancer cells and kill them, while passing up healthy cells. Immunotoxins are suicide nanorobots, designed to perform their single function one time, killing the target cell and being destroyed themselves in the process.

As you might imagine, the major challenge is finding just the right antibody. Cancer cells are very similar to normal cells, when looked at from the outside. Researchers have had good luck targeting a class of proteins that are involved in cell-cell interaction and adhesion. Cancer cells often display a different set of these molecules, providing cancer-specific targets for binding of the immunotoxin.

The choice of toxin is also important. Typically, the toxic component of a plant or bacterial toxin is used. These toxins are exquisitely effective: A single molecule can kill an entire cell. The reason for their effectiveness is that they are enzymes. They enter cells and then jump from one molecule to the next, inactivating each in turn. This is far more effective than poisons like cyanide and arsenic, which inactivate cellular molecules one-on-one, requiring millions of molecules to kill a cell.

Drugs May Be Delivered with Liposomes

For decades, researchers have explored the use of nature's own packaging system for custom delivery. Artificial membranes, under appropriate conditions, form small, closed vesicles called *liposomes*, composed of a lipid bilayer that encloses a small droplet of water. Methods are available for creating liposomes in a range of sizes, from about 20 nm to 10 μm. They are useful in many applications. Because they have a water droplet trapped inside, they can carry water-soluble cargo such as proteins, but because they are composed of lots of carbon-rich lipids, they can also carry water-insoluble cargo inserted into the membranes. They are at home in the human body, because they may be created with naturally occurring lipids. Liposomes are nontoxic, nonimmunogenic, and biodegradable.

Although liposomes have not turned out to be the answer to all drug delivery problems, they are attractive delivery mechanisms in a number of useful applications (Figure 6-5). They are particularly useful for delivering drugs that are very potent and very toxic. In addition, liposomes can increase the lifetime of drugs that are rapidly cleared from the blood. Liposomes are also particularly effective in some cases for targeting drugs to given locations, for instance, for delivering drugs to macrophages in the blood. These *immunoliposomes* are cleared from the blood by macrophages, which engulf and digest them, releasing the drugs directly inside the cell in the process. This is perfect for the treatment of *Leishmania* infection. *Leishmania* is an organism that infects macrophages and lives inside the cells. Because they are inside the cells, the immune system cannot control the infec-

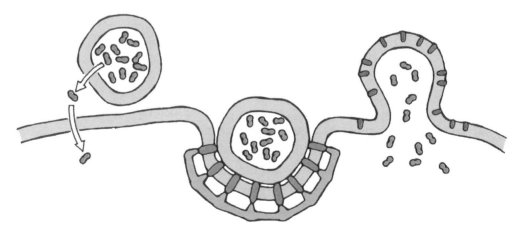

Figure 6-5 Liposomes interact with cells in several different ways. Some simply bump into the surfaces of cells and bind weakly, as shown on the left. Then any drugs carried inside the liposome can slowly leak out, providing a small but consistent dose to the local area. In other cases, the liposomes are drawn into cell by the normal endocytosis mechanisms, as shown in the center. As they pass through the lysosomes, where internalized molecules are digested, the liposome is degraded and the molecules inside are released. Finally, in special cases, liposomes can be designed to fuse with the cell surface, as shown on the right. They then dump their contents directly into the cell and incorporate their lipids directly into the cell membrane.

tion, but drugs in immunoliposomes can attack the problem at its location. Liposomes can also decrease the neurotoxicity of drugs. Many drugs pass out of the bloodstream into the brain. However, if a drug is stored in a large liposome, it is blocked by the blood-brain barrier, reducing possible side effects on nervous tissue.

Liposomes effectively penetrate the skin, delivering molecules to cells in the lower layers. This property has been used to great advantage in the cosmetics industry, where liposomes are added to skin creams and other products applied to the body surface. Liposomes may be designed to deliver specific lipids to the skin by including the lipids in the bilayer. Vitamins, tanning agents, sunscreens, and many other substances are also commonly enclosed within liposomes for delivery.

Because naked lipid vesicles are not normally found in the blood, liposomes are rapidly removed from the bloodstream. To increase their effective lifetime in circulation, researchers have developed *stealth liposomes* that evade the natural clearance mechanisms. One approach is to coat the liposomes with large, neutral polymers. These form a barrier that interacts only weakly with the natural molecules in the blood, such as the antibodies and blood-clotting proteins that normally interact with foreign bodies. These liposomes may circulate for several days in the bloodstream, acting as a time-release reservoir of drug molecules.

Artificial Blood Saves Lives

Countless lives have been saved by blood transfusions, but transfusions of whole blood have a few disadvantages. Whole blood has a short shelf life and must be carefully matched for blood type. Also, the possibility of viral contamination is always a cause of concern, even though HIV and hepatitis viruses are no longer a problem because of effective screening of blood. This has spurred research in creating artificial blood for use during surgery or in emergencies.

Purified hemoglobin has many potential advantages as a blood substitute. It may be used on any patient without the need for blood typing, because the blood type sugars are found on the surfaces of red blood cells, not on the hemoglobin molecules. Purified hemoglobin may be sterilized to remove pathogens, and it may be stored for more than a year. However, the first experiments showed that natural hemoglobin, by itself, is not a useful blood substitute. In 1937, it was found that transfusion of purified hemoglobin delivered oxygen successfully but was highly toxic to the kidneys. Hemoglobin is normally a tetramer but dissociates into dimers when free in the blood. These dimers are rapidly filtered by the kidney, where it accumulates to toxic concentrations.

Researchers are now looking to bionanotechnology for answers, engineering hemoglobin for use in blood replacement. The first approach was to create a hemoglobin complex that is stable in the blood. Purified hemoglobin was cross-linked, using small chemical agents like glutaraldehyde to connect lysine amino acids on the surface. Two forms were successful: polyhemoglobin, in which several hemoglobin molecules are linked into a larger complex, and cross-linked tetrameric hemoglobin, in which specific cross-links were created between subunits within the tetramer. A more targeted approach was taken with recombinant DNA methods. A modified hemoglobin was designed that contains two hemoglobin subunits fused into one chain (Figure 6-6). This then forms a stable complex similar to the tetramer, but with two of the subunits covalently linked together. In all of these cases, the resultant protein was large enough to resist filtering by the kidney but retained excellent oxygen-carrying capabilities.

Researchers have also explored encapsulation of hemoglobin within artificial containers, effectively building a custom, nonimmunogenic red blood cell. Hemoglobin has been successfully enclosed in liposomes. They are effective for delivery of oxygen, but they are rapidly removed from the blood by the natural systems that patrol for defective red blood cells. The life span has been improved by carefully tailoring the surface properties of the lipids, making them similar to the sugars that coat normal red blood cells, and creating liposomes that are smaller than normal cells.

Thomas M. S. Chang has extended this work beyond natural materials. He is exploring the use of biodegradable polymers for the creation of hemoglobin nanocapsules. Polymers such as polylactide and polyglycolactide are degraded in the body into water and carbon dioxide, making them a safe vehicle for encapsulation. They are also stronger and more porous than lipids, so less membrane material is needed and the hemoglobin content of the ma-

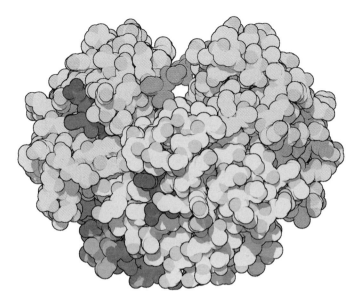

Figure 6-6   In this engineered hemoglobin, two of the subunits (shown in pink) are combined into one long chain. This stabilizes the entire complex, making it more suitable for use in blood replacement.

terial can be higher than with liposomes. As a final refinement, antioxidant enzymes such as superoxide dismutase, catalase, and metHb reductase may be included inside the capsule. Hemoglobin is reactive and continually generates toxic oxygen radicals. Natural red blood cells contain these additional enzymes and use them to destroy these dangerous products. Chang has designed polymer capsules that include these detoxifying enzymes along with hemoglobin, creating a safe and effective blood substitute.

Gene Therapy Will Correct Genetic Defects

With our growing understanding of the human genome, we now have the possibility of understanding and correcting specific genetic defects such as diabetes or cystic fibrosis at their source. Gene therapy is an area under intense research, with the hope that useful therapies will be produced in the near future. Tempering this excitement, researchers and policy makers have proceeded slowly. Because gene therapy seeks to make changes at the heart of our humanity, it raises profound ethical questions. Each generation will have to decide which changes are ethically acceptable.

Several approaches are being explored. The most obvious is to correct a missing or defective protein. The correct gene is added to the cell, where it then creates the active protein. This will provide a way to correct cells that have lost their ability to create insulin or an essential enzyme. One might also incorporate a new gene, with entirely new properties, to correct a problem. For instance, researchers are developing methods to incorporate toxic genes into cancer cells, so that the cancer cells make poisons and kill themselves.

We might also want to control a rogue molecule within a living cell. Antisense therapy attempts to correct this problem. A new gene is added that produces a therapeutic RNA that seeks out the messenger RNA for the rogue protein and binds to it, blocking production of the protein. Similarly, the instructions for specific ribozymes may be added. The ribozymes are then built inside the cell, where they attack the messenger RNA for the unwanted protein. In other approaches, the gene for an antibody is made inside the cell, to seek out the rogue protein and inactivate it.

The major challenge is finding safe and effective ways to deliver these new genes into cells, and in particular, into only the few cells that need the correction. The most common approach thus far is to use retroviruses. These viruses infect a cell and incorporate their DNA into the genome of the cell. This is exactly what is needed. Because the DNA is inserted into the cell's genome, it will reproduce faithfully when the cell divides. Unfortunately, retroviruses insert the DNA in random places, so they may disrupt normal genes when the new DNA is added. Adenovirus, a virus that causes mild coldlike symptoms, has also been tested. It will deliver DNA inside cells, but it does not insert the DNA into the genome. It is a temporary solution, and the engineered genes are only added for one generation of the cell.

Other approaches that do not use viruses are also in testing, although they are not as efficient and reliable as delivery with viruses. One method is incorporation of DNA into liposomes. The liposomes interact with the cell surface, transferring the DNA inside. Surprisingly, a naked circle of DNA by itself will also find its way into cells and direct the creation of the proteins that it encodes. These naked DNA preparations are becoming promising candidates for the design of vaccines. As with adenovirus, however, these DNA strands will not incorporate into the cell's genome, so the effects are temporary. Researchers are attempting to modify the therapeutic DNA, however, to create artificial chromosomes that will replicate faithfully along with the cell's genome when the cell divides.

General Medicine Is Changing into Personalized Medicine

Nanomedicine is changing the face of medicine, moving in small steps toward a new paradigm of *personalized medicine*. Based on the genetic makeup of each individual patient, therapies may be tailored to prescribe the most effective forms of treatment and minimize potential side effects with the particular variants of enzymes found in each person's cells.

Already, this approach is revolutionizing the treatment of HIV infection. As described above, HIV mutates rapidly, allowing rapid development of resistance to anti-HIV drugs. It is now possible to take a sample of a patient's blood and determine the viral subtypes that are most common. A drug regimen may then be designed to fight the particular collection of viral mutants that have arisen over the course of the infection. To make the technique widely effective, several improvements will be necessary. First, it is necessary to identify a wide variety of viral subtypes with only a small blood sample. Microchip technologies are perfect for this type of analysis, once the parameters of the search are defined. More problematic is the design of the treatment strategy. A wide variety of drugs are currently available for blocking many of the proteins in HIV. Accurate prediction techniques are needed to design the optimal therapy based on the population of viruses that are found in each patient. In principle, this type of analysis could be performed monthly, keeping therapy on top of any new resistance that develops.

A similar approach may be envisioned for cancer treatment. Cancer is the result of many different genetic changes that allow cancer cells to grow without control and to migrate to different parts of the body. By cataloging the specific changes found in each patient, targeted therapies can be designed to attack the problem directly.

SELF-ASSEMBLY AT MANY SCALES

Biology is filled with ideas and innovations that may be applied to engineering. A classic example is the design of airfoils based on the shapes of bird wings. The idea of self-assembly is the newest concept from biology to inspire engineers. Although used in cells primarily at the nanoscale, engineers have applied the concept at many levels, both larger and smaller than the biological examples.

Self-Assembling DNA Scaffolds Have Been Constructed

Through decades of research and testing, Nadrian Seeman has developed a bionanotechnology for creating defined multidimensional structures composed of DNA. Seeman sees several advantages in the use of DNA in nanotechnology. First, it is programmable and predictable. The base-pairing rules allow ready design of any possible pairing of strands. In Seeman's words, "the very properties that make DNA so effective as genetic material also make it an excellent molecule for programmed self-assembly." Second, DNA oligonucleotides of any desired sequence are readily synthesized by automated solid support techniques. Finally, DNA double helices are stiff polymers, at least in the range of a few turns of the double helix, or about 10 nm. Add to this the relative stability of DNA under physiological conditions and the many natural enzymes available for manipulating DNA, and you have an attractive nanoscale building material.

Seeman has pioneered the use of branched DNA structures to create multidimensional objects. Seeman takes a modular approach. The building blocks are designed from several strands that assemble to form a cross-shaped complex. The sequences of each leg of the cross must be unique to ensure that only one final structure is formed. Sticky ends are left on each branch and are used to link blocks together into larger structures. By matching the sequences of these sticky ends, assemblies of any shape may be designed (Figure 6-7).

These branched complexes have been used to create closed polyhedra, such as cubes and truncated octahedra, with DNA double helices at the edges of the structure. To make the construction of these objects more efficient, the growing structure is immobilized on a bead. When immobilized,

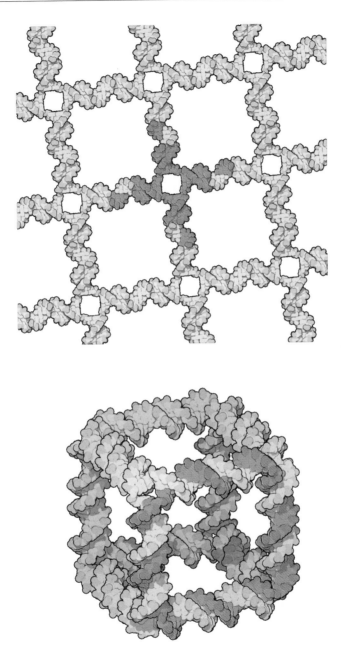

Figure 6-7 Large structures of DNA are built with modular subunits with sticky ends. A two-dimensional network is created with a four-armed module as shown at the top. A cube is created of eight three-armed modules as shown at the bottom. When DNA ligase is used to link the modules together, the structure is composed of eight topologically linked DNA circles.

each growing structure may be treated as a single object in isolation, so improper intermolecular interactions are reduced and side products are minimized. Also, the reaction sequence was designed such that each successful synthetic step produces a topologically linked product, so that a denaturation step could be used to remove improperly formed molecules.

The joints between the arms in these branched structures, unfortunately, are flexible, so the resultant polyhedral structures are not rigid. To create rigid structures, Seeman has employed a second key design for DNA nanotechnology: crossover structures. These are designed by arranging two or more helices side by side and then exchanging strands at the points where they touch. By appropriate choice of sequence, the strands will assemble in these side-by-side structures with all bases paired with their appropriate neighbors, forming a structure that mimics two helices rigidly locked together.

Seeman has created a set of modular tiles using these crossover molecules. Two or more types of tile are designed to create a two-dimensional lattice. These have been built and analyzed by atomic force microscopy. Perhaps the most exciting development has been the incorporation of short hairpin loops on the face of the lattice. These may be clipped by restriction enzymes, leaving a sticky end available for attaching to custom DNA components. In this way, the two-dimensional DNA lattice will act as a support, allowing addition and removal of components through base pairing. Seeman envisions the use of these lattices in nanoelectronics and for storage of data.

Cyclic Peptides Form Nanotubes

Reza Ghadiri has designed a modular concept for the self-assembly of nanotubes. He uses cyclic peptides that are perfectly designed to stack on top of one another. The peptide groups in his circular molecules have hydrogen bonding groups facing up and down from the plane of the circle, forming hydrogen bonds that glue one ring to its neighbors. The clever aspect of the design is in the side chains. Ghadiri builds these circles from amino acids with alternating handedness, so that the side chains all face radially out from the center of the circle, leaving a smooth channel through the center (Figure 6-8).

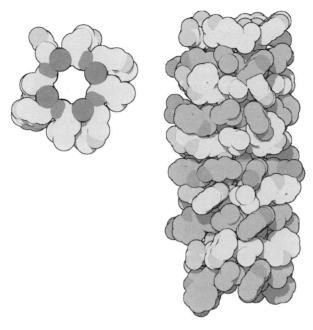

Figure 6-8   The small cyclic peptide shown on the left assembles into a nanotube by stacking one on top of the next. The ring of oxygen and nitrogen atoms, shown in bright red, form hydrogen bonds that link the rings together. Note how all of the side chains are pointed outward, leaving a smooth hole down the center.

When these rings are created from leucine and tryptophan amino acids, the resultant nanotubes have a carbon-rich outer surface. These rings assemble into nanotubes that span lipid membranes. By choice of the size of the rings, the diameter of the channel can be tuned. A circle of eight amino acids forms a channel of about 0.45-nm diameter. When synthesized, these formed effective channels for metal ions but blocked passage of larger molecules. A slightly larger ring of 10 amino acids has a channel of about 0.9 nm. Those nanotubes have been shown to allow the passage of glucose.

However, when the surface is designed with a hydrophobic stripe of amino acids on one side and a stripe of hydrophilic amino acids on the other, the rings assemble into rafts that float in the surface of the membrane, half immersed and half exposed to solvent. These nanotubes disrupt the orderly structure of membrane and can break cell membranes. A very similar approach is taken by the natural protein mellitin, which is found in bee venom. By careful choice of amino acids, Ghadiri hopes to target these attackers to specific cell types.

Fusion Proteins Self-Assemble into Extended Structures

Proteins are the preeminent raw material for creating large structures. Looking to nature, we can find all shapes of filaments, cups, cages, and lattices composed of protein subunits. The different shapes are specified by the relative orientation of combining sites on the surface of the protein, which define the geometry of interaction of the subunits. Todd Yeates and coworkers have developed a general method to create our own nanoassemblies composed of protein subunits.

They start with proteins that naturally form low-order assemblies, such as dimers or trimers. Then they fuse two of these proteins together back-to-back, leaving the natural combining sites exposed. When mixed together, these fusion proteins link up one to the next using their natural combining sites. By appropriate choice of the linker connecting the two proteins, many different types of assemblies may be created. They have reported two successful examples. A fused protein composed of bromoperoxidase (which normally forms a trimer) with influenza M1 matrix protein (which forms a dimer) assembles into a tetrahedral cage. A fused protein composed of the M1 protein and carboxyesterase (which also forms a dimer) assembles end to end to form a long filament.

Small Organic Molecules Self-Assemble into Large Structures

Researchers have been inspired by the process of self-assembly in biomolecules and have begun to explore self-assembly at other scale levels. Chemists are looking to self-assembly as a path to creating large assemblies from small organic molecules. This is currently a vigorous field of research, producing many clever approaches and applications.

One of the early successes was the creation of molecular capsules that capture given molecules. These are needed for chemical separations and other specific recognition applications. Charles Pederson, Jean-Marie Lehn, and Donald Cram were awarded the Nobel Prize in 1987 for their work in creating complexes of small molecules formed by self-assembly. Pederson began this work with the study of *crown ethers,* cyclic molecules that have several oxygen atoms pointed toward the hole in the center. By changing the size of these molecular rings, different metal ions are specifically recognized and bound inside. Building on this concept, chemists have designed countless *host* molecules, each with a specific cavity inside to hold a particular *guest* molecule. These include cup-shaped *calixarenes* with specific groups around the rim that guide binding of guest molecules and *carcerands,* composed of two cups that entrap a guest molecule in between (Figure 6-9).

Taking the lead from DNA base pairing, George Whitesides has explored complexes of melamine and cyanuric acid that form extended structures. These are planar organic molecules that have hydrogen-bonding groups around their perimeter. These associate side by side using hydrogen-bonding networks reminiscent of base interactions in DNA, and they stack on top of one another like typical aromatic compounds. By themselves, these would form infinite complexes, but by careful design specific structures may be built. Whitesides directs the assembly with two methods. First, he uses preorganization, by specifically tethering two or more rings

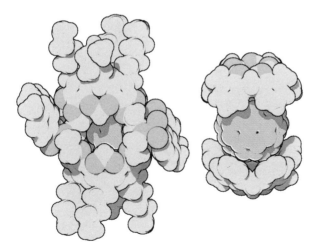

Figure 6-9   Calixarene molecules are designed to create a container for a guest molecule. On the left is a large calixarene that holds methoxytoluene, and two calixarenes bound to a buckminsterfullerene are shown on the right.

together in the desired orientation. Second, he adds large substituents in strategic positions on the rings, which crowd out potential neighbors.

Molecular self-assembly is also being used to create molecular assemblies with moving parts. Fraser Stoddart has created a collection of molecules termed *rotoxanes*. These are composed of two parts: a molecular ring and a molecular rod. The molecule is self-assembled so that the rod passes through the hole in the ring. Then it is capped at either end so that the ring cannot escape (Figure 6-10). These types of molecules are being explored, among other things, as molecule-sized rotary motors and as abacus-style computational memory.

Larger Objects May Be Self-Assembled

George Whitesides has pioneered the design of micrometer- to centimeter-scale objects that self-assemble into defined larger structures. We might ask,

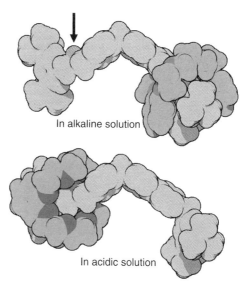

In alkaline solution

In acidic solution

Figure 6-10 The circular ring of this rotoxane molecule changes position based on the pH of the surrounding solution. The ring, shown in gray, is free to move along the center of the rod, shown in pink, but is blocked at either end by large groups that keep it threaded on the rod. A key nitrogen atom, shown with the arrow, changes between charged and neutral states as the pH changes, alternately pushing the ring back and forth along the rod.

what are the advantages of self-assembly at a level where direct manipulation is possible? He describes a few. Self-assembly is intrinsically parallel, allowing the simultaneous assembly of a large collection of subunits. Self-assembly allows submicrometer accuracy in positioning, which may be a problem with traditional industrial assembly methods. In addition, self-assembly is relatively insensitive to errors in registration of components.

His early experiments explored two-dimensional self-assembly. The objects were small hexagons, a few millimeters in dimension, composed of poly(dimethylsiloxane), a hydrophobic compound. Some edges of the hexagons were made hydrophilic by oxidation. The patterned hexagons, with hydrophobic and hydrophilic edges, were then floated at the interface between water and perfluorodecalin. The system was then agitated, and hexagons associated through their hydrophobic edges. As with biological self-assembly, these interactions are reversible, so eventually imperfect complexes changed into the assembly with the most stable arrangement of subunits. These experiments showed that, like biological systems, the form of the final complex was highly dependent on the shape and directionality of the interaction between subunits. He also demonstrated self-assembly with even smaller objects, with dimensions of about 10 μm. These subunits were constructed with a combination of photolithography, electrochemistry, and metal evaporation techniques. By proper patterning of the edges and faces of small hexagonal subunits, large crystalline arrays were self-assembled.

Subsequent work in Whitesides's laboratory has explored a number of exciting variations on this process. By careful design of interfaces that bond to themselves but do not bond to other interfaces, he has created hierarchical complexes composed of two or more differently shaped subunits. A modification of this method was used to create an analog of DNA information storage and readout. Three-dimensional structures have also been created with similar methods, by using polyhedra for the subunits and agitating them in solution instead of at a liquid interface.

With a view toward applications, Whitesides and his coworkers have used these methods to create three-dimensional, self-assembled electrical networks (Figure 6-11). The individual subunits are polyhedra. Some of the faces carry electrical components, such as light-emitting diodes, and the remaining faces carry contact pads that interact with other polyhedra. The

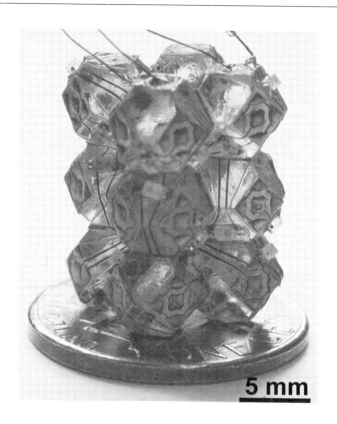

Figure 6-11 Whitesides constructed this three-dimensional electrical circuit by using self-assembly. Each component contains a series of contact pads, wires, and light-emitting diodes. To assemble the circuit, the contact pads were coated with low-temperature solder and the components were suspended in warm water. When gently agitated, the components with matching patterns of contacts self-assemble into the final circuit. (Figure 3c from Gracias, D.H., Tien, J., Breen, T.L., Hsu, C. and Whitesides, G.M. (2000) "Forming Electrical Networks in Three-Dimensions by Self-Assembly." *Science* 289, p. 1171.)

pads are patterned to interact specifically with pads of other subunits. Instead of hydrophobic interactions, the pads are coated with a low-temperature solder. The subunits are suspended in warm water, which melts the solder and allows the subunits to self-assemble by coalescing of solder on the pads.

HARNESSING MOLECULAR MOTORS

When we think of machines in our familiar world, we think of machines with moving parts. Motion requires motors. So, quite naturally, an early goal of nanotechnology has been the harnessing of powered motion at the nanoscale. In bionanotechnology, several spectacular demonstrations have already been achieved. Thus far, the harnessing of biomolecular motors has been used primarily for the study of the motors themselves. However, the techniques are general and applicable to nanoengineering applications.

ATP Synthase Is Used as a Rotary Motor

The rotation of ATP synthase was controversial when proposed by Paul Boyer, but recent nanoengineering experiments have demonstrated the rotary motion directly by immobilizing the motor and connecting an object that is large enough to observe microscopically. Researchers at the Tokyo Institute of Technology first demonstrated the rotation of ATP synthase in 1997. They engineered a form of the motor with 10 extra histidine amino acids at one end, which extend from each β-subunit. These short tails have a high affinity for nickel ions, so a glass coverslip was prepared with another enzyme, horseradish peroxidase, with nickel bound. This effectively glued the motor to the coverslip with the axle pointing up. To attach a microscopically visible load, the axle was engineered to include a new cysteine residue. Biotin was then added to the cysteine, providing a ready handle for the protein streptavidin. Finally, an engineered version of actin, with a biotin attached, was also connected to the streptavidin. Rotation was observed in an epifluorescence microscope when ATP was added to the surrounding solution. They have improved their techniques in the subsequent years, using a smaller bead as the load and resolving two separate 30° and 90° rotations for each ATP consumed.

Researchers at Cornell University have made a first step toward combining bionanotechnology and MEMS. They have constructed a nanopropeller out of inorganic materials and attached it to ATP synthase (Figure 6-12). The trick is finding a way to connect the biological components to the inorganic components. As with other work on ATP synthase, the tight association of histidines with nickel was used as the glue. First, a support for the

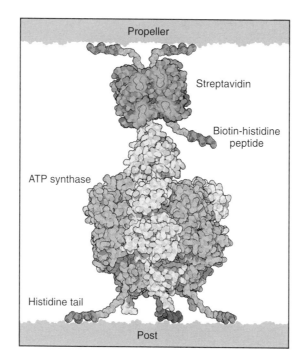

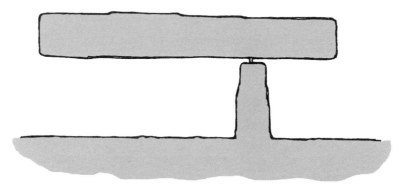

Figure 6-12 ATP synthase has been engineered to turn a nanopropeller. The setup is shown at the bottom. A series of posts were created on a flat substrate with electron-beam lithography and then coated with nickel. Small propellers were created by similar techniques. The biological motor component, tiny in size compared to the post and propeller, is assembled in between. The motor component is shown at the top. ATP synthase was engineered with three histidine-rich tails that bind to the post. The rotor of ATP synthase was engineered with a biotin, which binds to streptavidin. The remaining three binding sites of streptavidin were filled with a short biotin-histidine peptide, which binds to the nickel surface of the propeller.

machine was created by electron-beam lithography. It was composed of an array of nanoscale posts, 50–120 nm in diameter and 200 nm tall. These were coated with nickel to provide a platform for attachment of the motor. Then an engineered version of ATP synthase similar to that described above was created, with histidine-rich tails on the motor domains and a biotin molecule attached to the axle. The tails attached the motor to the posts. Finally, the nanopropeller was attached to the axle of the motor. The propeller was a rod 150 nm in diameter and 740–1400 nm in length, coated with nickel. A linker molecule was created, composed of a histidine-rich peptide and biotin. The peptide bound to the propeller, and streptavidin was used to link the biotin on the axle with the biotin on the nanopropeller. Of the 400 posts created on the chip, 5 sites were seen to rotate when ATP was added. The other propellers were completely immobile, with no random fluctuations, so the researchers suggest that they are attached at multiple sites, gluing them in place. The working propellers turned for several hours before the propellers broke loose.

Molecular Machines Have Been Built of DNA

The first moving machines designed from scratch have been built with DNA. As described above, DNA is an attractive building material, because the pairing rules for complementary strands allow defined topologies to be designed. Apart from allowing defined construction, the binding and unbinding of strands may also be used to do work.

Researchers at Bell Laboratories have created a pair of tweezers composed of DNA (Figure 6-13). The concept is direct: A hinged molecule is created, double stranded in the center with a gap to make it bendable, and with single-stranded extensions at each end. Then another DNA strand, termed the *set* strand, is designed that matches the two extensions. When added to the solution, the set strand pairs with the two ends, bringing them together and closing the tweezers. To allow opening of the tweezers, the set strand is designed so that it is too long, leaving some of its bases unpaired. A competing *fuel* strand is then added, which is designed to pair with the entire length of the set strand. It pulls the set strand off the tweezer, forming a long double-stranded helix that is discarded. Given the strength of base-

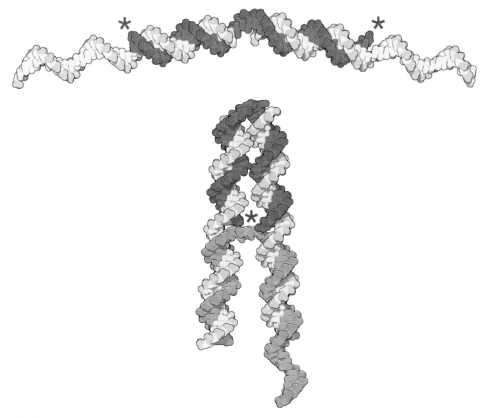

Figure 6-13 DNA tweezers contain a flexible hinge in the center and single-stranded regions at either end, as shown at the top. When the pink strand is added, it pairs with the two ends, closing the tweezers. Researchers have attached various molecules at the two sites marked by asterisks, demonstrating that they are brought together when the tweezers close.

pairing interactions, the authors predict that the tweezers would have roughly the force of a kinesin or myosin motor.

Seeman and his coworkers have created a similar mechanical device that rotates. Again, a mechanism is created from two DNA strands, this time, a complex arrangement of hairpins and double strands. Two set strands are then added, which lock the mechanism into one state. The state is released by the addition of fuel strands that remove the set strands, leaving the device in an unstructured state. Addition of a second type of set strands locks the mechanism into the second rotated state. A second set of fuel strands then release this state. By cycling through the strands, the machine can be made to oscillate back and forth.

DNA COMPUTERS

Molecular computers have been a dream, and a goal, of nanotechnology from its inception. The possibility was introduced by Richard Feynman in his seminal lecture, and numerous schemes have been proposed ever since. The first successful man-made molecular computers have taken their lead from biology, using biology's premier information-carrying molecule.

DNA is used widely in natural biological systems as a data storage medium, but quite surprisingly, it is used in only limited ways for computation inside cells. The regulation of gene expression is a form of computation, where various promoter, repressor, and enhancer proteins bind to the DNA and their effects are summed to determine whether a particular gene will be expressed. This computation, however, is relatively wasteful of resources. A new protein is made for nearly every new logical operation, and the space within the DNA genome that is used for computation, encoding these proteins and providing binding sites for them, may be very large.

Researchers have taken a different approach to DNA computation. Instead of using a network of proteins that perform a computation by binding along a DNA strand, researchers have designed computers that use the pairing of DNA strands as the mode for computation. DNA has many attractive properties that make it excellent for the design of molecular computers, foremost of which is sequence-specific interaction of DNA strands. Under appropriate conditions, single strands of DNA will pair specifically with their exact complement, even when it is part of a mixture of trillions of competing sequences with very similar sequence. Also, individual DNA strands may be selected and copied with the polymerase chain reaction (PCR) to give quantities of material that are sufficient for sequencing or testing.

Staggeringly large computations may be performed with DNA computers. A typical simulation may sift through 10^{20} different potential solutions to a problem, selecting the single one that is correct. Because of the small size of the individual molecules, this requires only a manageable amount of material. The computation is also massively parallel. Once the computation is set up, all of the possible solutions are examined simultaneously. Finally, because each component is molecule-sized, the computations are cheap and energy efficient. Critics have warned, however, that as the problems are scaled larger, the details of manipulating large quantities of DNA will become a problem.

DNA computing also has several potential disadvantages. Perhaps the greatest limitation is that the error rates in copying reactions, although reduced to very small levels by evolution, may still be too great for many large-scale applications. This is an intrinsic property of the pairing of DNA, so there are no ready solutions. It is worth noting, however, that biology has taken advantage of this small error rate to allow evolutionary optimization, and we may be able to turn it to our own advantage. In addition, most current DNA computers are not reusable, requiring synthesis of a new set of molecules for each new problem. Recent work has suggested some new avenues toward design of renewable methods, but thus far each new application has required a radically different design for the DNA computer. This, however, is not a surprise, given that the field is so new.

Reports of DNA computation have shown great cleverness in designing methods to translate the algorithm into the physical pairing of DNA strands. There is also a healthy skepticism, both from critics and from the researchers themselves, about the potential of DNA computation. In particular, questions have been voiced about its scaling to problems that rival those currently solvable by electronic computers. Many see DNA computation as the first step toward more robust molecular computation designs. In the words of Leonard M. Adleman, "They enlighten us about alternatives to electronic computers and studying them may ultimately lead us to the true 'computer of the future.'"

The First DNA Computer Solved a Traveling Salesman Problem

In 1994, Adleman first demonstrated that DNA could be used to create a molecular computer. In his experiment, he solved a "traveling salesman problem" (Figure 6-14). The problem is to find a path within a given graph

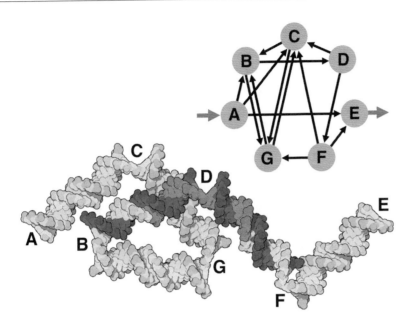

Figure 6-14 DNA computation may be used to solve a traveling salesman problem. The graph is shown at the top. We are looking for a path that starts at node A, passes through each node only once, and exits at node E. Adleman created 7 unique DNA strands that correspond to each node of the graph, shown in pink in the lower illustration, and 14 strands that correspond to each path from node to node, shown in gray here. The molecule at the bottom, formed by spontaneous pairing of these strands, represents the solution to the problem.

that visits each node only once and begins and ends at defined spots. These types of problems become exponentially more difficult as the number of nodes are increased. Adleman designed a DNA ligation system that tests all possible routes through the graph and provides an answer.

He created two types of single-stranded DNA, each 20 nucleotides long. The first type corresponds to each allowed path through a node, with 10 base pairs representing the incoming path and 10 base pairs representing the outgoing path. A different nucleotide sequence was chosen for each, to ensure that each path was represented uniquely. The second type of DNA was created to correspond to each of the paths between nodes. These are designed to match with the ends of the node DNA segments. In Adleman's experiment, the graph included 7 nodes and 14 allowed paths between nodes.

All of the DNA strands were then combined and allowed to anneal. The gaps in the double strands were then ligated. At this point, the mixture of DNA contains many long double-stranded helices representing every possible path through the graph. In his experiment, he included about 3×10^{13} copies of each piece, so there was more than enough to represent all possible paths through the graph.

The trick is then to isolate the one DNA helix that corresponds to the answer. Adleman subjected the mixture of ligated double helices to three screens. The first was a PCR amplification using primers that correspond to sequences at the desired beginning and end of the path. This amplifies only those DNA helices that begin and end at the proper node. Then he used gel electrophoresis to screen for size, picking only those that were the proper length to visit exactly seven nodes. Finally, he did an affinity screen, checking the strands for the presence of sequences that correspond to each node. Only those strands that contained all of the nodes were retained.

Satisfiability Problems Are Solved by DNA Computing

Satisfiability problems have become a benchmark for the testing of new DNA computing methods. These problems attempt to solve a Boolean logic equation. They are computationally difficult, requiring exponentially growing computational resources as the problems increase in size. The problem is to take a logic statement such as:

(a and b)

where a and b can be true or false, and devise values for the parameters that will yield a true outcome. In this simple case, both a and b must be true to give a true outcome. The typical problem is composed of a series of clauses composed of parameters and "or" operators, such as:

(a or b)

which are then connected together by "and" operators:

(a or b) and (a or c)

These problems have a particularly straightforward implementation in DNA computation.

To solve satisfiability problems, a long DNA strand is created that includes a segment representing each parameter in the equation. For the simple equation above, the DNA strand would have three segments, one for a, one for b, and one for c. Then, two unique DNA sequences are chosen for each parameter, one for true and one for false. A complete set of possible answers is then created, by creating all possible DNA strands composed of either true or false sequences at each parameter position. Our simple problem would have a true-true-true strand (representing parameters a-b-c), a true-true-false strand, and all the other combinations. All of these DNA strands are then mixed together.

Then the trick is to remove all of the wrong answers, leaving the proper answers in the test tube. Adleman has devised a renewable method for performing this step. A gel electrophoresis column is created that contains complementary DNA strands attached to the gel. For the first clause, the column contains strand complementary for "a is true" and "b is true" and the column retains any DNA strands that contain these values, letting all of the "a is false" and "b is false" strands pass through. Then the strands are released and subjected to a second column that tests for the second clause. The strands that remain bound are the answer to this simple problem.

For more complex problems, the researcher simply scales up the DNA strands to include more parameters and uses more successive columns to include more clauses in the equation being solved. Adleman reports solution of a problem with 20 parameters and 24 clauses in the equation, successfully separating the single correct solution from over a million possible incorrect answers.

A Turing Machine Has Been Built with DNA

Much of contemporary computation is based on the concept of a Turing machine—a machine that reads a list of data and performs operations based on the values at each point along the list. For example, a very simple machine might be designed to tell whether there is an even or an odd number of values in a list. This machine would have two states, S0 and S1. As the machine reads down the list of information, it has two operations that modify the state: The first changes S0 to S1, and the second changes S1 to S0. The machine performs a simple operation, determining whether there is an even or an odd number of data points in the list. The machine reads down the list of data: If it ends up in S0, there was an even number of values, and if it ends up in S1, there was an odd number.

Ehud Shapiro and coworkers have presented a DNA-based Turing machine that can perform this computation. The machine is composed of a DNA strand that contains the list of instructions, a collection of DNA strands that encode the allowed operations, and two enzymes that perform the computation. The computation is performed by snipping off data one digit at a time, using the location of the cut to encode the state of the machine. They have implemented several simple algorithms, demonstrating the potential power of the approach.

MOLECULAR DESIGN USING BIOLOGICAL SELECTION

Biology does not *design* new machinery to fulfill its functional needs—instead, biology *discovers* new machines. Natural evolution discovers new nanomachines by selecting the desired function from a large library of different machines. A large pool of variants is produced by random mutation, and the best examples are selected. Successive rounds of mutation and selection lead to better and better machines. Two elements are necessary for this process: a method for creating the library of variants and a method for selecting and reproducing the most functional individuals in this library. Cells create their library by random mutation of the genome and do the selection and reproduction by expressing the genome into proteins and selecting individuals that are more fit and able to create offspring.

Researchers have harnessed the evolutionary paradigm for design and used it to create functional nanomachines in the laboratory. The three methods described in this section use an approach modeled after biological processes, but selecting functional molecules in weeks instead of the millennia required by evolution. All three methods search a large library of different variants for the individuals that show the greatest function of interest. All rely on biological processes to provide selection and reproduction of the best individuals from this large library.

The molecules designed by these evolution-inspired methods have great potential for use in nanomedicine, as biosensors, and in custom chemical synthesis. All three of these techniques are well established for use in biomolecular research, for probing functional aspects of natural biomolecules. More recently, researchers are applying these techniques to nanotechnology, harnessing the principles of biology in yet another way to reach our own technological goals.

Antibodies May Be Turned into Enzymes

Based on a clever concept, antibodies have been used to create customized enzymes that catalyze reactions of interest. As described in Chapter 5, enzymes often catalyze a chemical reaction by stabilizing the transition state, the unstable molecular structure on the reaction path between the reactants and the products. Remarkably, an antibody that binds to this same transition state will often act as an enzyme. This selection is conveniently accomplished by injecting a molecule that mimics this transition state into an animal and gathering the antibodies that are formed. The trick is finding the proper mimic. We need a molecule that is similar in shape and chemical character to the transition state but is stable under the conditions of antibody selection.

This is a powerful approach, because the immune system prefabricates an enormous library of different antibodies. This provides a diverse collection of potential active sites with different chemical and physical characteristics from which to select candidates. Of course, this process requires understanding of the reaction mechanism so that effective transition state mimics can be designed. Proof of the concept was provided in 1986 with an ester-cleaving antibody developed at the Scripps Research Institute and an antibody that cleaves a small carbonate developed at the University of California at Berkeley. In the years since then, the approach has been remarkably successful.

By careful design of the transition state mimic, antibodies catalyzing many different types of reactions have been obtained. Catalytic antibodies that cleave esters, obtained with mimics that contain a phosphonate group that is similar to the tetrahedral carbon transition state, have been among the most successful. They perform exactly as expected. They display kinetics similar to those seen in natural enzymes, they show specificity for molecules similar to the mimic used to select the antibodies, and they accelerate the rate of the reaction by up to a millionfold. Other cleavage and carbon-transfer reactions have been reported with similar transition state mimic molecules.

Some chemical transformations are of interest to chemists but are not normally performed in biological systems. Diels–Alder cycloaddition reactions are a perfect example. They are widely used in organic chemistry for the construction of six-membered rings, but natural enzymes typically take different approaches to creation of rings. With bridged compounds that mimic the transition state, catalytic antibodies performing many variants of the Diels–Alder reaction have been found (Figure 6-15). Of particular interest is a class of reactions that are normally disfavored under normal chemical conditions. These include reactions that have two or more possible transition states, leading to several different products. Normally, the reaction would proceed primarily to products with the transition state of lowest energy. However, the other products may be favored by using a catalytic antibody that specifically binds to the less favorable transition states.

Catalytic antibodies currently suffer from some limitations. Most catalytic antibodies have relatively low activity and limited turnover. These limitations are due in part to limitations in the evolution possible in laboratory selections. Natural enzymes have been optimized over far larger numbers of generations. Approximately 1000 antibody-secreting hybridoma cell lines can be obtained through monoclonal antibody techniques, and a million or more members may be constructed from antibody fragments. However, only a fraction of these are typically screened in a typical catalytic antibody experiment. Improved experimental methods will allow greater numbers of different antibodies to be screened, allowing selection of greater functional properties.

The use of transition state analogs to screen for catalytic antibodies is also a severe limitation, contributing to low activity and low turnover number. Natural enzymes are optimized for action directly on the actual reaction and optimized for proper release of products after the reaction is completed, allowing catalysis of many molecules. In some cases, timely release of products may be promoted by designing a chemical or conformational change into the product, so that it ultimately adopts a form that is quickly released.

Researchers are currently looking to biology for more ideas as to how to proceed. By specific design of different molecules to select antibodies, cat-

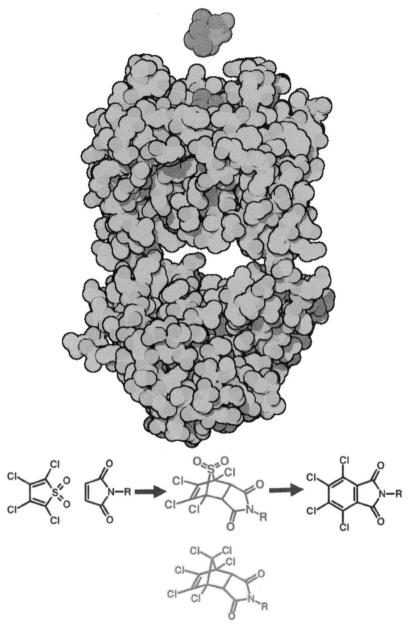

Figure 6-15 A catalytic antibody that catalyzes a Diels–Alder condensation reaction (shown at the bottom) was developed with a transition state mimic, shown in red.

alytic antibodies are being produced that use cofactors, such as vitamins, inorganic ions, and metal ions. This opens up the possibility of reactions that require chemical characteristics other than those provided by the 20 natural amino acids. Site-directed mutagenesis and modification of amino acids in the antibody after protein synthesis are also being used to refine the function of existing catalytic antibodies.

Catalytic antibodies are showing great promise as general tools for organic synthesis. For instance, a catalytic antibody was used at the Technion–Israel Institute of Technology to perform a key step in the synthesis of (−)-α-multistriatin, a pheromone of the European elm bark beetle. This antibody allowed creation of key chiral centers in the compound. As current limitations of efficiency and cost are being addressed, catalytic antibodies are emerging as essential tools for the synthesis of organic compounds with specific handedness.

Catalytic antibodies are also being tested for applications in nanomedicine. They are ideal for performing specific tasks in the human body because they are nonimmunogenic and will not elicit a response from the immune system. There is also the exciting possibility that patients could be immunized with the transition state mimic itself, creating the necessary catalytic antibodies in the patient's own body by using the normal capability of the immune system. If this concept turns out to be practical, we could use these molecules to induce catalytic antibodies in the way that we use vaccines to prime the immune system for future attack by viruses.

One example under testing is the creation of antibodies that activate drugs on site. To reduce side effects, anticancer drugs are often created as inactive *prodrugs* that are activated only when they encounter cancer cells. The current approach is to link an antibody that targets the cancer cell to an enzyme that activates the drug. The advantage of using a catalytic antibody as the activating enzyme is the wide choices available. A prodrug may be chosen that is resistant to all natural enzymes in the patient's body and is only activated by the catalytic antibody.

In a second application, catalytic antibodies that inactivate recreational drugs are being developed. Antibodies against drugs have been previously explored for use in treatment of drug addiction. However, these antibodies bind one-to-one with drug molecules, and the supply in the blood is rapidly exhausted by the first dose of drug, leaving the body susceptible to future doses. Catalytic antibodies against the drug remain active as long as they are designed with sufficient turnover rate. For instance, a catalytic antibody that cleaves cocaine has been obtained at the Columbia University College. It shows a sufficient rate to cleave all of a typical dose of cocaine before the drug has time to cross the blood-brain barrier to its site of action.

Peptides May Be Screened with Bacteriophage Display Libraries

With efficient synthetic techniques, it is currently possible to create a vast library of random peptides. By running this library over a given protein target, we could isolate the peptides that bind to the protein. The problem, however, is analyzing the peptides that we isolate—there would be such small quantities in the library that analysis would be difficult. Bacteriophage display libraries are a solution to this problem. They allow screening of a large library of random peptides or proteins for function. The trick is the use of *bacteriophages*, viruses that attack bacteria. The bacteriophages are engineered so that they contain both the peptide or protein that is being tested and the DNA that specifies the amino acid sequence of the protein. The researcher simply selects the single bacteriophage that makes the protein with the desired function, grows the phage in large quantities, and then looks at its genome to determine the amino acid sequence of the protein.

Typically a large library of recombinant bacteriophages is constructed. A short peptide or an entire protein is grafted onto a protein that is found on the end of the bacteriophage. Bacteriophages are engineered with many variants of the peptide or many variants of a short cassette within the protein. For instance, common bacteriophage display libraries include all possible combinations of amino acids in a peptide about five amino acids long. When this library is grown in bacteria, bacteriophages are produced with each of the different sequences displayed on their surfaces and the DNA encoding the particular sequence stored inside.

The library is then screened with an appropriate target. For instance, if we are looking for a peptide that binds to a given protein we can mix the protein with the library and look for bacteriophages that stick to it. A typical library may include up to a trillion different peptides, but only a small fraction bind specifically to the target. Then these specific bacteriophages are isolated and reproduced in bacteria. After several rounds of screening, a small number of specific-binding peptides may be obtained. The DNA in the bacteriophage is then isolated to determine the amino acid sequence of the specific peptide that it encodes.

Similarly, if we are searching for an enzyme that will bind to a given small molecule we can create a library with the whole protein attached to the bacteriophage coat protein. The library will include modified versions that include all possible modifications of amino acids within small cassettes inside the protein. The library is then screened for binding to the desired target, selecting the best bacteriophages for further analysis.

There are several design issues when creating a library. First, the displayed peptide or protein must be accessible for testing and it should be in a native conformation. In successful libraries, peptides are often flanked with a constant sequence of amino acids that provide a linker separating the peptide from the bacteriophage surface. This makes it more accessible and reduces interference from the rest of the bacteriophage.

The library of randomized peptides is actually obtained by creating a library of randomized segments in the bacteriophage DNA. When these are transfected into bacteria, they form the library of peptides on the bacteriophages. A common approach is to use a "NNK" motif to specify each amino acid codon, where N is any nucleotide and K is either G or T (or G or C). Limiting the nucleotides used in the last position reduces codon bias (some amino acids are specified by many codons, some by only one) and reduces the number of stop codons that are included in the randomized sequence. More elaborate schemes have also been developed for doping the randomized peptides with given sets of amino acids.

Bacteriophage display libraries are powerful tools for discovery of novel biomolecular functionalities. For instance, Angela M. Belcher has used bacteriophage display to discover an interface between bionanotechnology and semiconductor technology. She has selected peptides that bind to specific faces of gallium-arsenide semiconductor crystals. Selection methods were designed to isolate specific peptides that bind, for instance, to one specific face of gallium-arsenide semiconductors but not to silicon. These may provide ready handles for interfacing inorganic components with bionanotechnology. By analogy with biomineralization, these peptides may also provide tools for patterning the formation of semiconductor crystals or other inorganic components.

Nucleic Acids with Novel Functions May Be Selected

With the discovery of natural *ribozymes*—RNA molecules with catalytic activities—and the recent discovery that the ribosome uses its RNA in the catalytic reaction, the possibilities of DNA and RNA as functional molecules are becoming clear. So, naturally, researchers have searched for methods to find RNA molecules with novel functions. SELEX (Systematic Evolution of Ligands by EXponential enrichment) has been the most successful approach to this goal. It mimics the process of natural evolution to identify oligonucleotides with specific functional properties. The process and its variants have been used to identify *aptamers*, nucleic acids that specifically bind to ligands, as well as custom ribozymes (Figure 6-16).

SELEX techniques isolate functional nucleic acids from a large library of random sequences in much the same way as bacteriophage display libraries are searched for functional peptides. A library of short DNA sequences is synthesized with a randomized segment. The DNA is then transcribed into a library of RNA molecules with RNA polymerase. These are subjected to one of a number of methods for selecting a given function, such as binding to a ligand or performing a chemical reaction, and the functional sequences are isolated. These sequences are then amplified by reverse transcription into DNA and then amplification with the polymerase chain reaction. The process is then repeated for multiple cycles, selecting sequences with greater and greater functional properties as the pool of sequences compete at each step. After 5 to 10 rounds, the pool is dominated by a few types of molecules that bind to tightly to the target.

A typical selection experiment is capable of selecting a single active molecule from a pool of over 10^{15} different sequences. For sequences of about 15 randomized nucleotides this size of pool provides coverage of all possible combinations, but if longer sequences are randomized, the experiment will provide only partial coverage of the total possible sequences. Surprisingly, this does not appear to be a major hindrance. In one study, researchers used selection experiments to find aptamers that bind to six

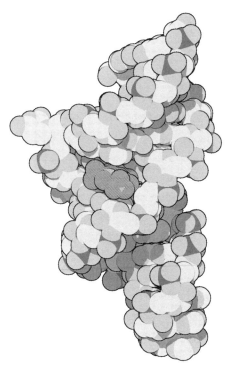

Figure 6-16 This RNA aptamer, discovered by SELEX methods, binds to the small organic molecule theophylline, shown in pink.

different organic dyes. They used a randomized sequence 100 nucleotides long and found that about 1 sequence in 10^{10} can form a specific binding site.

The key to the process is the design of an effective selection method. For design of aptamers that bind to proteins, the library is mixed with the protein target and the complexes are trapped on a nitrocellulose filter. The nonfunctional RNA is washed through the filter, and the functional RNA is then eluted from the protein bound to the filter. For aptamers that bind to small molecules, the target ligands may be tethered to agarose or another insoluble substrate and the selection can be performed on a chromatographic column. Again, the RNA library is washed over the agarose-ligand substrate and only functional RNA molecules bind.

A variety of aptamers that bind to small molecules have been discovered by these methods, with targets including nucleotides and amino acids, cofactors, and antibiotics. Aptamers typically bind to small molecules with micromolar dissociation constants, with occasional nanomolar dissociation constants for particularly favorable interactions. Thus they rival the binding characteristics of proteins, despite the fact that they do not have the chemical diversity afforded by the 20 natural amino acids.

Aptamers tend to be unstructured when free in solution. They adopt a defined structure as they fold around the ligand. The binding sites are generally formed deep within the folded nucleic acid chain and are often composed of loops rich in adenine and guanine, which form unusual base pairs to stabilize the structure. Binding strength and specificity are provided by a combination of the stacking of base rings on hydrophobic and aromatic groups in the ligand, specific hydrogen bonds and electrostatic interactions, and a form-fitting shape. In general, tightest binding is obtained with ligands that contain planar aromatic groups along with hydrogen bonding or charged groups. As with recognition by proteins, nonplanar carbon-rich molecules, which lack convenient molecular handles, are the most difficult for recognition by aptamers. Recognition is quite sensitive in some cases. For example, an aptamer has been isolated that can distinguish between caffeine and theophylline, which differ by a single methyl group.

Aptamers that bind to proteins, protein assemblies, or even whole cells have also been discovered. Proteins provide a larger and more varied surface than small ligands, making them excellent targets for aptamer selection. The first experiments showed surprisingly successful results. RNA aptamers that bind to the blood-clotting protein thrombin were selected from a pool of sequences with 60 randomized nucleotides. About 1 in 100,000 sequences bound to thrombin with nanomolar dissociation constants. Many subsequent experiments have shown that specific aptamers may be routinely isolated for binding to nearly any protein.

By designing an appropriate selection technique, ribozymes that cleave or ligate nucleic acid chains have also been obtained. For instance, to select a ribozyme that cleaves nucleic acids, a pool of nucleic acid sequences is created with both the site of cleavage and the randomized ribozyme test sequence together in a single strand. The strand is attached to a solid support through the end closest to the cleavage site. In some cases, the randomized sequence forms an active ribozyme, cleaving the strand and releasing the portion of the strand with the functional site. Only functional sequences are released and are easily isolated in the solution. To isolate a ribozyme RNA ligase, a slightly different approach is effective. This approach relies on connection of two pieces, which must be present together for successful selection. Ribozymes that catalyze alkylations and Diels–Alder reactions have been isolated with a similar selection based on successful covalent bonding of two separate molecules with attached tags.

By using an approach similar to the selection of catalytic antibodies, *in vitro* selection can be used to discover all types of catalytic ribozymes or deoxyribozymes. Nucleic acids are selected that bind to transition state analogs of the desired reaction intermediate. Thus far, this approach has not yielded the diversity of successes shown with catalytic antibodies, but several functional ribozymes have been discovered. For instance, a ribozyme that catalyzes the addition of metals to porphyrin rings has been selected. The proposed mechanism of action involves a deformation of the planar ring, favoring the distorted conformation of the metal complex.

Remarkably, ribozymes that are controlled by allosteric motions have been isolated. These ribozymes are created by linking two functional RNA molecules: One binds a regulatory molecule, and the other performs the catalysis. Successful examples have been isolated where the hybrid molecule shows cleavage or ligase function that is regulated by binding of nucleotides or other small molecules. This is a remarkable example of engineering both the function and the control mechanism of a bionanomachine, setting an important precedent for bionanotechnology.

Stability is a significant problem with RNA aptamers, particularly when they are used in nanomedicine. Blood contains efficient enzymes for degrading foreign nucleic acids. Several approaches to improve stability have been tried. The free hydroxyl group may be modified with amino or fluoro groups, or the SELEX method may be modified to yield aptamers composed of nucleotides with the opposite handedness. In both cases, the modified nucleic acids are resistant to degradation. Of course, these modifications must be incorporated during the selection, to ensure that they do not destroy the binding capability of the aptamer.

Functional Bionanomachines Are Surprisingly Common

Work on RNA aptamers and ribozymes has revealed that functional molecules are more common than one might predict. Aptamer selections have shown that approximately 1 in 10^{10} molecules will bind specifically to a given small molecule. Protein-specific aptamers are even more common: Approximately 1 in 100,000 will have the desired functional characteristics. Active ribozymes are somewhat rarer: Experiments with RNA ligase ribozymes have shown frequencies of about 1 in 10^{13}, but they are still far from unique in the library of possible sequences. As reported by Wilson and Szostak, the likelihood of finding functional sequences is increased for several reasons. First, many positions in an aptamer or ribozyme are not important for activity, but merely provide infrastructure. Second, there are many distinct sequences that adopt the same structure, such as a structural helix that may be formed by different sequences. Finally, there are many distinct structures that satisfy the same selection pressure for a given function. These principles are also true of proteins but have not been explored as quantitatively because of the relative ease provided by RNA selection experiments. The common occurrence of functional nucleic acids (and presumably proteins) is a boon for nanotechnology, showing that both specific binding and catalysis are achievable in synthetic molecules at rates and specificities rivaling evolutionarily optimized natural biomolecules.

ARTIFICIAL LIFE

Surprisingly, one of the most complex possibilities of bionanotechnology, both technically and ethically, is the application with the longest history of research and success. Creation of life has been an area of active research since the beginnings of molecular biology. Most of the work has focused on uncovering how life started on Earth. These experiments have attempted to create everything from the building blocks of life to functioning organisms by using only the materials and natural processes available in the early history of Earth. Today, bionanotechnology opens the possibility of creating new life forms with all the properties of natural cells.

Of course, when seeking to create life, we are not limited to biological materials, or to physical materials at all. Thus far, most researchers building artificial cells have stayed close to the biological design, given that we are still struggling to understand the existing mechanisms that create natural life. Computer science, however, has ranged much further, employing key biological concepts to improve computational function. Examples include evolutionary approaches to computational searching, neural networks, computer viruses, and the elusive goal of artificial intelligence. The concept of life has invigorated computational science and, in the future, will open new avenues for design and creation in the physical world. Thus far, many of the pieces needed to create artificial cells have been studied, including self-reproducing containment structures, self-replicating informational molecules, and methods for powering these processes.

Artificial Protocells Reproduce by Budding

In the 1960s and 1970s, two laboratories made the first key steps toward creation of a basic infrastructure for artificial life. A. I. Oparin in Moscow and Sidney W. Fox at the University of Miami created systems that resemble simple cells and perform some of the basic functions of life, such as enzymatic synthesis and reproduction. Oparin studied *coacervates,* colloidal droplets that are formed when two different macromolecules, such as a protein and a carbohydrate, are mixed in solution (Figure 6-17). These particles are cell-sized, ranging from a micrometer to half a millimeter, and often show a thickened "membrane" boundary at the surface. Oparin found that if enzymes were added to the mixture, they were incorporated into the droplets and remained active. His laboratory tested a series of *protocells* that contained key metabolic functions. For instance, if phosphorylase enzymes were added to the droplets, they created starch when fed a solution of activated glucose. The drops grew as starch built up inside and eventually budded into smaller daughter droplets. The process, however, slowed over time as the limited amount of phosphorylase enzyme was separated into the daughter droplets, reducing its concentration. Similar protocells were created with a variety of biological processes, including RNA synthesis and oxidation/reduction reactions.

Sidney Fox explored the creation of primordial proteins. He found that

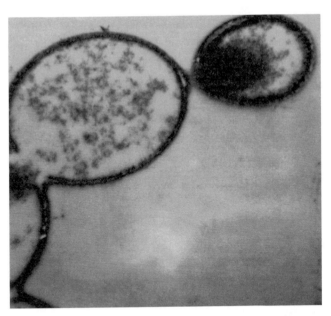

Figure 6-17 Coacervate droplets form spontaneously from mixed solutions of biological molecules. These coacervates, created by A. I. Oparin, are formed of protamine and polyadenylic acid. They grow and bud in ways that resemble living cells. (Figure taken from Dickerson R.E. (1978) "Chemical Evolution and the Origin of Life." *Scientific American* 239(3), p. 83.)

when a dry mixture of amino acids was heated to about 130°C, they spontaneously formed chains of about 200 amino acids in length. He termed these *thermal proteinoids*. When then heated in solution, these proteinoids formed cell-like droplets a few micrometers in diameter, complete with a cell-like outer membrane. As with Oparin's coacervates, these proteinoid spheres grew in size over time and budded into daughter spheres when they got too large. He found that these proteinoids themselves also had primordial enzymatic action. For instance, they could polymerize ATP into short oligonucleotide strands.

These experiments showed how simple biomolecules, similar to those that could have been created in the early history of life, can provide the infrastructure needed for an effective artificial cell. Today, liposomes are the more common approach for containment. Artificial protocells that microencapsulate any combination of enzymes may be conveniently created by stirring the lipid components with the molecules to be enclosed. Numerous variations have been tested, creating membranes with different levels of permeability and enclosing all manner of enzymes, multienzyme systems, magnetic materials and even entire natural cells.

By careful design, these lipid-based enclosures can be coaxed into re-producing. Pier L. Luisi and coworkers have developed a system of self-re-producing vesicles. They start with a solution of an insoluble precursor of a fatty acid, such as oleic anhydride. This oily substance separates when added to water solutions. When oleic anhydride is layered onto an alkaline water solution, the anhydrides are slowly broken down into oleic acid, which forms vesicles in the water solution. The process starts out slowly, because the anhydride oil has only a small surface of contact with the water solution. However, after vesicles begin to form, anhydride molecules rapid-ly bind in the surfaces of the vesicles, where they are rapidly cleaved into the fatty acid. The vesicles multiply rapidly as the existing vesicles catalyze the production of more.

Self-Replicating Molecules Are an Elusive Goal

Self-replication of information-carrying molecules is a key functionality needed for artificial life. In natural cells, a multicomponent system is used for this task—in fact, one might think of the entire protein synthesis ma-chinery (or as Dawkins has proposed, an entire organism) as being simply a method to replicate DNA. Creation of a single molecule that can self-repli-cate has been an area of intense, but elusive, research. Most research has fo-cused on RNA and RNA-like molecules, trying to create simple systems where new strands are created by using existing strands as a template. These systems are designed to test possible hypotheses about the origin of life on Earth.

In the 1980s, L. E. Orgel and coworkers showed that activated nu-cleotides, with reactive groups added to the phosphates, spontaneously as-semble along RNA templates and form complementary strands. In some cases, this assembly was quite efficient. For example, a strand of guanine formed spontaneously with a cytosine strand as a template. However, when mixed sequences were tried, complications arose because of the different reaction rates, different association affinities, and possible misalignment of templates and product strands.

More recently, the search has turned to ribozymes. Using powerful *in vitro* selection techniques like those described above, workers at the Whitehead Institute have developed RNA molecules that can catalyze the extension of RNA strands based on a template. The ribozyme can add about 14 new nucleotides to the strand and gets the pairing correct about 97% of the time. This reaction is not processive, however, so it is not possible for the ribozyme to create a copy of itself. It is a remarkable first step, pointing the way toward design of a self-replicating molecule.

ATP Is Made with an Artificial Photosynthetic Liposome

Artificial organisms will require a source of chemical energy. Researchers at Arizona State University have looked to natural photosynthesis for inspiration and have created a simplified light-powered system that creates ATP with remarkable efficiency (Figure 6-18). Their clever approach creates an artificial mimic of the light-capturing mechanism found in natural photosynthetic reaction centers. They created a three-part molecule. At the center is a porphyrin that absorbs light, similar to the less stable chlorophyll molecules used by nature. At one side, they connected a naphthoquinone. Soon after absorbing a photon, the porphyrin transfers an electron to this quinone. On the other side, they attached a carotenoid. This carotenoid then donates an electron back to the porphyrin. This creates an excited state with an electron on the quinone and a positive charge a long distance away on the carotenoid. Normally, this decays in less than a nanosecond back to the ground state. But, quite remarkably, they have designed a system to capture the excited state, just as in natural photosynthesis, using it to create ATP.

As in plant cells, they used the excited electron to create an electrochemical proton gradient. The mimic molecule is embedded in liposomes, with all of the carotenoid ends facing in. Then a lipid-soluble quinone is added to the mix. The excited electron is transferred to this quinone on the outer surface of the membrane. It then quickly picks up a proton from the surrounding solution, creating a neutral form. This diffuses through the

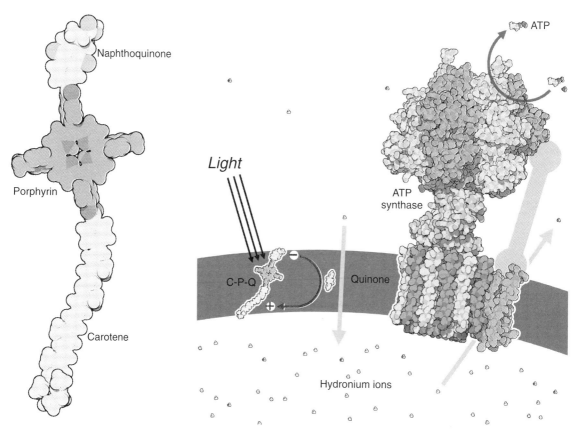

Figure 6-18 An artificial photosynthetic molecule was created by linking a porphyrin, which absorbs light, with a naphthoquinone and a carotene, which allow the transport of the electron that is activated by light. Researchers placed this molecule in a liposome along with a lipid-soluble quinone and ATP synthase. The result was a liposome that could use light to build ATP.

membrane and, on the inside, it donates its electron back to the carotenoid, releasing the proton inside the liposome. The result is a pumping of protons into the liposome, powered by light. They then used this gradient to power ATP production. They added ATP synthase to the membrane, which produced ATP as protons were allowed to flow outward. A remarkable yield was observed despite the extreme economy of the molecular design. One ATP was formed for every 14 photons of incident light.

Poliovirus Has Been Created with Only a Genetic Blueprint

The genomes of many organisms have been sequenced, revealing the basic plans for life. This raises an important question: Can we create a living organism based only on these genetic plans? Jeronimo Cello, Aniko Paul ,and Eckard Wimmer have made the first step. They have taken genetic information available in public databases and built a working virus from scratch. This is a remarkable achievement, indicating that the actual physical structures are the only pieces needed to achieve the biological function. As the authors relate: "If the ability to replicate is an attribute of life, then poliovirus is a chemical [$C_{332,652}H_{492,388}N_{98,245}O_{131,196}P_{7501}S_{2340}$] with a life cycle."

Poliovirus is a simple virus, composed of a single strand of RNA 7741 bases long and 60 copies of each of four proteins. To create an infectious virus, these pieces needed to be constructed and then assembled. They began with the nucleotide sequence of the virus, which is available in public databases. Then they purchased pieces of DNA from commercial sources that covered the entire sequence of the virus and assembled them into a double-stranded DNA version of the viral genome. The viral RNA was then created from this template with a purified RNA polymerase enzyme. This RNA was then added to a cellular extract, which contained only the cytoplasm of a human cell. This extract contained all of the necessary machinery for protein synthesis and quickly constructed the viral proteins from the synthetic RNA. Natural proteases in the extract also processed the viral protein properly, forming the four separate poliovirus proteins. The parts then assembled spontaneously into infectious viruses (Figure 6-19).

By all expectations, creation of a living cell based only on its genome will be far more difficult. Much of the mechanism of life is encapsulated in the structure of the cell, which must be passed from parent to offspring. For instance, there is no indication that the shape of mitochondria, the double-layer form of the nuclear membrane, or the details of the other membrane-bounded compartments in higher cells are encoded in the DNA. These morphologic details are passed directly from parent to offspring by example, not through instructions. As our understanding of the functioning of the genome deepens, we are certain to find many aspects of cells that are passed *epigenetically* rather than through the genome. When engineering ar-

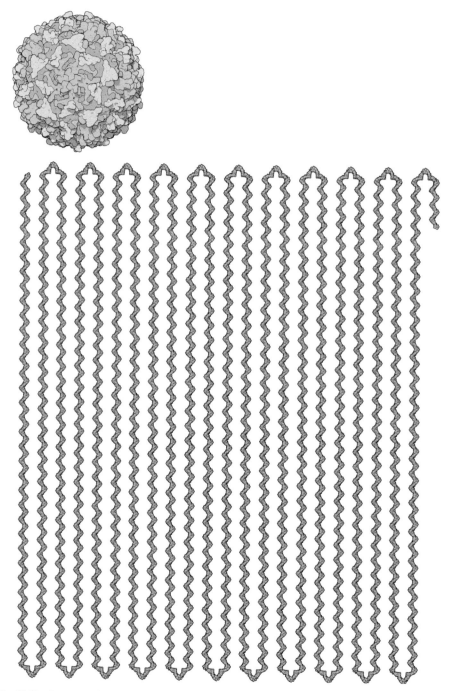

Figure 6-19 Poliovirus, synthesized entirely from scratch for the first time in 2002, is composed of a long RNA strand carried inside a hollow protein capsid.

tificial cells, these same features will have to built directly into the constructs, instead of letting the genome control all of the work.

HYBRID MATERIALS

When we think of technology, we think of automobiles, cellular phones, computers, and other durable products of mechanical and electronic engineering. Abundant technologies are available, ranging from the micron level up, that harness diverse materials, from glass, plastic, and metal, to lubricating oils and flexible rubber, to conductors, semiconductors, and superconductors. Interfacing these technologies with the more delicate materials of bionanotechnology is yielding some exciting early fruit.

Nanoscale Conductive Metal Wires May Be Constructed with DNA

One approach to nanocomputing is to create traditional electronic devices miniaturized to molecular dimensions. Nanoscale components are beyond the resolution of top-down lithographic approaches, but the assembly of electronic components with biological self-assembly as a template is already showing success. Some of the first applications have used DNA as a template for the construction of nanoscale wires. In several cases, researchers have taken advantage of the many negative charges on the DNA phosphates to guide the deposition of narrow tracks of inorganic material. Early examples include the formation of cadmium sulfide nanoparticles along DNA strands and the patterning of fullerenes that are derivatized with a positively charged amino group.

Researchers at the Technion–Israel Institute of Technology have constructed nanoscale wires composed of silver grains deposited along a strand of DNA. The process involved several steps. First, small DNA oligonucleotides were attached to two gold electrodes. To do this, a disulfide group was added to the end of the strands. The sulfur atoms then form a tight interaction with the gold electrodes. The template DNA strand was then created with sticky ends that pair with the two oligonucleotides. The long template strand was attached, bridging the two electrodes by anneal-ing to the two end connectors. To create the wire, a concentrated solution of silver nitrate was then added, which replaces the sodium ions that normally associate with the DNA phosphates. This forms a line of silver ions along the DNA backbone. Each of these silver ions was then used to nucleate larger silver aggregates, which were then used to create silver grains in a process akin to photographic development. The resultant wire was 12 μm long and about 100 nm wide.

As a further refinement, they used recA protein, a recombination protein that pairs DNA strands with similar sequences, to protect portions of the DNA strand. This forms a removable "resist" that blocks the deposition of metal and creates functional gaps in the nanowire. They term this process *molecular lithography*. It requires several steps. Aldehyde groups are added to the DNA, and the complex with recA is formed, creating the template. Silver aggregates are then formed by the aldehyde groups on the DNA, in portions where it is exposed. Finally, the silver aggregates serve as catalysts for the deposition of gold, creating conductive gold wires (Figure 6-20).

Patterned Aggregates of Gold Nanoparticles Are Formed with DNA

Chad A. Mirkin has used DNA to create defined aggregates of gold nanoparticles. Short single strands of DNA are modified with propylthiol

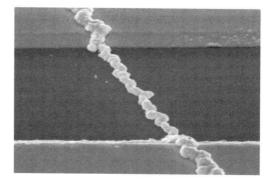

Figure 6-20 DNA has been used as a template to create this gold wire connecting two electrodes. (Figure 2a from Keren, K. et al. (2002) "Sequence Specific Molecular Lithography on Single DNA Molecules." *Science* 297, p. 73.)

groups, which bind tightly to the surface of the gold nanoparticles. A DNA strand with sticky ends is then added to the solution. It pairs with the nanoparticle-bound sequences and positions two neighboring particles at a defined distance from each other, defined by the length of the connecting DNA strand. As Mirkin says, this scheme "allows one to control particle periodicity, inter-particle distance, strength of the particle interconnects, and size and chemical identity of the particles in the targeted macroscopic structure." By employing the complex branched and three-dimensional DNA structures that have been developed by Seeman, it is easy to imagine a variety of complex nanoparticulate constructions.

One application has already emerged from this work. When the gold nanoparticles aggregate, the color of the solution changes from red to blue. This has been used to create a colorimetric assay for specific DNA sequences. Gold nanoparticles are coated with single-stranded DNA that is complementary to the two halves of the DNA sequence to be assayed. If the DNA is present in test solutions, it will aggregate these particles, inducing a color shift that is easily detected.

DNA Flexes a Sensitive Mechanical Lever

Many different types of molecules bind onto surfaces and induce surface stresses. If this surface is on a thin sheet of material, the molecules can cause the sheet to bend when they attach. Researchers at IBM and the University of Basel have harnessed this property to create a microscale silicon cantilever that detects DNA (Figure 6-21).

The cantilevers were constructed of silicon by microfabrication techniques. They are 1 μm thick, 100 μm wide, and 500 μm long, coated with gold on one side. Short DNA single strands, 12 to 16 nucleotides in length, were synthesized with a sulfur atom at one end, which fixed them to the gold surface. The complementary DNA strands were then added to the solution, and the bending was observed by watching the deflection of an optical beam bounced off the cantilever. The method was surprisingly sensitive. They were able to identify the difference in binding between the perfectly matched DNA and DNA strands that had a single mismatched base pair.

This is a potentially useful technology because the DNA that is being tested can be used directly with no modifications. Other methods require

Figure 6-21 A microfabricated silicon cantilever has been used to detect specific DNA sequences in a sample. The sample DNA binds to oligonucleotides bound to the surface of the cantilever, inducing stresses on the surface that cause it to bend. (Figure 1 from Fritz, J. et al. (2000) "Translating Biomolecular Recognition into Nanomechanics." *Science* 288, p. 316.)

labeling of DNA samples with radioactive isotopes or other molecules that are easily detected. In this system, only the native DNA strand is needed. The device is also reusable. The DNA sample can be removed, allowing use on other samples. The small size of the cantilevers also allows the creation of parallel arrays, allowing large numbers of different sequences to be tested on one small chip. The researchers propose that the magnitude of the deflection is sufficient to operate microfluidic devices, allowing a direct connection between a molecular event—DNA binding—and a microscale operation.

Researchers Are Harnessing Biomineralization

Strong, resilient biomaterials may be created by using a combination of biological materials and minerals. As described in Chapter 5, many different examples of composites of minerals and biomaterials may be found in nature. Researchers are beginning to harness these natural materials to enhance the properties of nanostructured materials.

Researchers can now create glassy amorphous silica on demand, using biological processes. In one application, researchers have looked to sea sponges for methods. These sponges build spicules composed of a thin strand of protein surrounded by silica. The proteins are termed silicateins, and it was found that they can catalyze the formation of amorphous silica by using tetraethoxysilane as a starting material. Looking to simplify the process, several simple polypeptides were tested for similar activity. It was found that large polymers, composed of blocks of 10 to 30 cysteine amino acids separated by blocks of several hundred lysine amino acids, could perform the same reaction. The blocks of cysteine align the tetraethoxysilane molecules and catalyze the polymerization, and the lysine amino acids are needed to make the entire chain soluble. Remarkably, different forms of silica could be formed under different conditions (Figure 6-22). In the absence of air, hard, transparent spheres were formed. However, in the presence of oxygen, the cysteine amino acids form disulfide linkages. When these were used to form silica, packed columns of silica were formed.

Looking to diatoms for inspiration, researchers at the Air Force Research Laboratory and the University of Cincinnati have isolated a short peptide that can catalyze the formation of silica from silicic acid (Figure 6-23). The 19-amino acid peptide has many serine amino acids and can perform this reaction under very mild conditions, forming silica nanospheres within minutes. They mixed this peptide with acrylates, which may then be photopolymerized with infrared lasers. A thin layer of the acrylate-peptide mixture was plated onto glass, and then a holographic image was used to cure the mixture. Areas of the pattern that received the largest doses of light

Figure 6-22 Two different morphologies of silica were formed with large peptides composed of cysteine and lysine. When sheltered from air, hard, transparent spheres were formed, but in the presence of oxygen, packed columns were produced. (Figure 2 from Cha, J. N. et al. (2000) "Biomimetic Synthesis of Ordered Silica Structures Mediated by Block Copolymers." *Nature* 403, p. 291.)

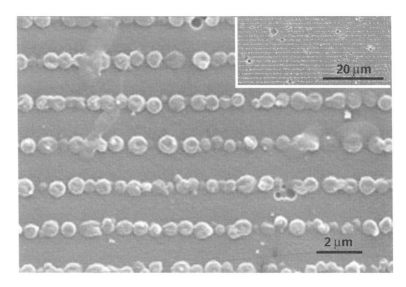

Figure 6-23 An ordered two-dimensional array of silica nanospheres was formed within in a hologram by using biomineralization peptides from diatoms. (Figure 3b from Brott, L. L. et al. (2001) "Ultrafast Holographic Nanopatterning of Biocatalytically Formed Silica." *Nature* 413, p. 293.)

are highly cross-linked, so the peptide tends to separate into the regions that receive lower doses. The template was then used to create silica nanospheres, which aligned in the grooves between the highly cross-linked areas. When tested, the silica-deposited hologram diffracted light 50 times more strongly than holograms without the silica nanospheres.

BIOSENSORS

Many applications in medicine, environmental analysis, and the chemical industry require sensitive methods for sensing small organic molecules. Our sense of smell and taste are designed to perform exactly this task, and the immune system recognizes millions of different molecules. Recognition of small molecules is a specialty of biomolecules, so they provide an attractive approach to the creation of specific sensors. Two components are needed: the recognition element and some mechanism for readout once the recognition element has found its target. Often, the recognition element may be taken unchanged from the biological source. The challenge is to design a suitable interface to a macroscale readout device.

Antibodies Are Widely Used as Biosensors

Antibodies are nature's premier biosensors, so it comes as no surprise that the development of diagnostic tests using antibodies has been one of the major successes of biotechnology. Perhaps the most familiar example is the simple test used to determine blood type. This is the simplest possible form of immunotesting, taking advantage of two properties of antibodies: their specificity and their ability to cross-link targets. The blood type test is composed of a collection of antibodies that recognize specific sugars on the surfaces of red blood cells. The antibody is added to the blood, and if the particular blood type is present on the cells, the antibodies bind to the surface, sticking many cells together. The result is a clumping of cells that is easily seen with the naked eye.

These types of precipitation reactions require large quantities of both the sample and the antibodies. Methods that are far more sensitive have been developed to allow smaller samples to be tested. Some of these are so sensitive that they can detect less than a thousand individual molecules in a sample. These tests attach a reporter group to the antibody, which creates the signal that is actually detected. Antibodies may be labeled with radioactive iodine or tritium and the presence of radiation used to quantify the amount of bound antibody. Powerful biosensors are created by linking antibodies to specific enzymes such as β-D-galactosidase or alkaline phosphatase. These enzymes then convert dye molecules to colored forms that can be detected. The most sensitive methods employ the detection of luminescent or fluorescent molecules, either connected to the antibodies or created by antibody-linked enzymes.

Pregnancy tests provide an example of how these tests can be streamlined for use in the home. Many variations are available from different providers using monoclonal antibodies that detect the presence of chorionic gonadotropin (CG), a small protein in the urine. A test from Abbott Laboratories uses a clever one-step process. The sample of urine is applied at one end, and it soaks through a fiber pad from one end to the other. First, the sample encounters a section with antibodies that have colloidal selenium particles attached, which are bright red. If the sample contains CG, it binds to the antibody. Then, as the sample continues through the pad, it drags the colored antibody and the bound CG with it. They next encounter two stripes of antibodies that are attached to the pad. One stripe is horizontal and contains specific antibodies that attach to the test antibodies. Some of the red antibodies stick here, creating at least a "minus" sign in all tests. The other stripe is vertical, and contains more antibodies that are specific for CG. If the sample contains CG, the CG-antibody complexes bind to these, forming a vertical stripe and completing a "plus" sign.

Biosensors Detect Glucose Levels for Management of Diabetes

Glucose biosensors are some of the most successful biosensors on the market today. People living with diabetes require convenient methods for monitoring glucose levels. Implanted sensors and noninvasive sensors are under development, but currently the most accessible approach is a handheld biosensor that analyzes a drop of blood.

The biosensor relies on the fungal enzyme glucose oxidase, which combines glucose and oxygen to form gluconic acid and hydrogen peroxide. A sensor can be designed to detect the amount of hydrogen peroxide formed. In the 1960s, Leland C. Clark had the clever idea to hold the enzyme very close to a platinum electrode with a membrane, so that the chemical changes could be followed by watching changes in the current at the electrode. This idea proved effective, and a series of laboratory-sized instruments were developed based on the sensing of peroxide.

To make a portable, consistent glucose biosensor for the home, however, a change in methodology was needed. The oxygen-to-peroxide change is hard to standardize, because of the need for consistent oxygen levels and interference by other molecules in the blood. Instead, a slightly different method was developed. Instead of oxygen, a mediator molecule is used to deliver the signal to the electrode. Ferrocene, a small molecule with an iron ion trapped between two cyclopentadiene rings, was found to be an effective mediator. Handheld meters that use disposable electrodes with enzyme and mediator are available commercially. Now, in a matter of seconds, glucose levels may be measured in a small drop of blood.

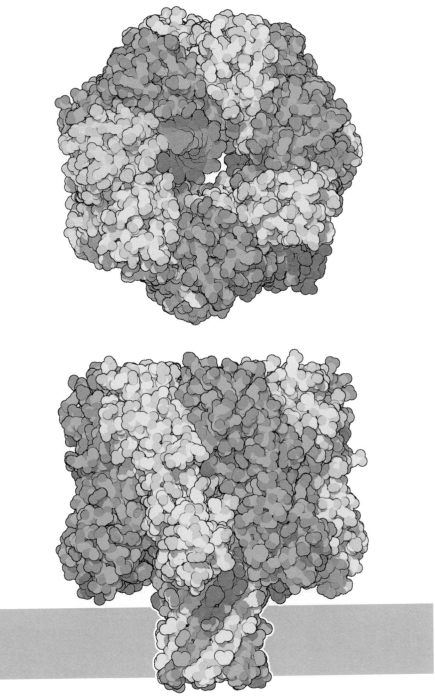

Figure 6-24 A DNA detector is created by tethering a short single strand of DNA (shown in pink) inside the pore of hemolysin (shown in gray). When the complementary strand is added to the solution, it binds to the tethered strand, blocking the pore.

A small modification can change a biosensor into a biofuel cell. In biofuel cells, specific enzymes are tethered to two electrodes, performing reactions that strip electrons from the fuel at one electrode and replace them on oxygen at the other electrode. Adam Heller has created a biofuel cell that uses glucose as its fuel. On one electrode, glucose oxidase extracts electrons from glucose, converting it to glucolactone. Then a second electrode is added with an enzyme that replaces the electrons, forming a closed circuit. For instance, the enzyme laccase may be used, which adds electrons to oxygen, forming water.

Engineered Nanopores Detect Specific DNA Sequences

Researchers at Texas A&M University have designed a biosensor that can detect short strands of DNA, about seven nucleotides in length (Figure 6-24). The sensor is based on the bacterial protein *hemolysin*. Hemolysin is composed of seven protein chains that form a pore through lipid bilayers. In nature, this is used as a toxin. As a biosensor, hemolysin is embedded in a membrane separating two chambers. An electrical potential is applied across the membrane, which draws ions through the pore from one chamber to the other. The current through this pore is monitored, and when the nanopores are blocked an abrupt change in the current is easily detected.

Hemolysin has a large chamber at one end, 3–4 nm in diameter, and a narrow tube that crosses the membrane, about 1.4 nm in diameter. To create the sensor, the researchers tethered a short DNA strand to one protein subunit inside the large chamber. This single strand does not block the pore, so ions are free to pass. The DNA strands to be tested are added to the solution, where they are drawn into the pore by the electrical potential. If a DNA strand does not match the DNA tethered inside, it passes quickly through the pore, reducing the current for about a tenth of a microsecond. If a DNA strand is complementary, however, it binds to the tethered strand and partially blocks the entry to the pore, causing a reduction in the current that lasts about 45 μs. Eventually, the strand dissociates and passes through the pore, restoring the current. By monitoring the time that DNA strands remained bound to the sensor, they could discriminate perfect matches from matches with a single mismatched nucleotide.

THE FUTURE OF BIONANOTECHNOLOGY

7

What would happen if we could arrange atoms one by one
the way we want them?

—Richard Feynman

What will the nanotechnology of the twenty-first century look like? Many people envision a technology of macroscopic machines shrunken to nanometer size: nanorobots with nanometer-scale gears, pulleys, gates, and latches; assemblers with rigid rectilinear struts and circular bearings; storage tanks surrounded by rigid walls of diamond. These machines emulate in every detail the machines that we use today in the macroscopic world. It is a compelling vision, filled with exciting prospects. But, is it the only course?

Instead, will nanotechnology be a forest of trees, capturing light to create plastic building materials, ceramic components, or entire dwellings? Will nanotechnology look like a stagnant pool, where cell-like nanomachines feverishly create tailored medicinal compounds, packaging them in custom delivery vessels? Nanotechnology may be a computer that runs not on electricity but instead on sugar and oxygen. Or nanotechnology may look exactly like a virus, but a virus customized to seek out and destroy cancerous tissues in each patient.

Bionanotechnology is a reality today. Through a confluence of experi-

Bionanotechnology: Lessons from Nature. By David S. Goodsell
ISBN 0-471-41719-X   Copyright © 2004 John Wiley & Sons, Inc.

mental knowledge from biology, chemistry, physics, and computer science, we understand the processes of life in sufficient detail to harness biomolecules for our own use. An entirely new realm of *wet-ware,* nanoscale machines that operate under physiological conditions, is open for the taking. Many of the goals of bionanotechnology and nanomedicine may be described as augmented biology: We are looking for nanomachines to perform functions normally done by biomolecules or by entire cells, but to do these jobs even better. Wet-ware is perfect for these applications, because these jobs will be performed in environments that are consistent with biological molecules—warm, wet, and salty.

Looking forward, the possibilities, some speculative science and some still the realm of science fiction, are tremendous. We have barely scratched the surface.

A TIMETABLE FOR BIONANOTECHNOLOGY

What might we expect in the future? Of course, it is always dangerous to speculate, because unforeseen developments are at the heart of most cultural and scientific shifts. Automobiles, trains, and airplanes are steadily shrinking and linking the world. The discovery of microscopic life and the subsequent international effort in antisepsis have doubled the length of our lives. Computers have made entire worlds of inquiry possible and raised important questions about our own minds. The world wide web has expanded our ideas of information and communication. In each case, a scientific or engineering advance opened a previously unimagined world. That said, what might we expect of bionanotechnology given foreseeable advances in our current understanding?

First of all, we can expect a solution to the protein folding problem. This will allow prediction of structure and function for a protein of arbitrary sequence, allowing the design of novel bionanomachines. This is a key step in bionanotechnology—extrapolating from existing machinery can only take us so far. By current expectations, effective computational methods for protein structure prediction should be expected in the next decade or so. And once natural proteins are understood, we can move on to the larger problem of improving and expanding the natural building materials for increasingly nonbiological applications.

Cellular engineering is another probability. Given the rapidly growing number of genomes and proteomes, we will have a full parts list of living cells in the near future. The coming decades will yield an understanding of how these parts are arranged and how they interact to perform the processes of life. With this understanding will come the ability to modify, tailoring new cells for custom applications. Already, bacterial cells have been engineered to create specific products, such as growth hormone and insulin. In the future, we will see cells that clean pollution, that make plastic and other raw materials, that fight disease, and that are used for countless other applications.

Organismic engineering will open many doors, but so far only natural methods are available. We have a long history of artificially accelerating evolution. People have bred organisms for centuries, creating organisms better fit to human welfare. With understanding of the molecular processes of development, the exciting possibility of engineering organisms from scratch will become possible. By directly modifying the genome of organisms, all manner of changes might be made. Already, genetic engineering is improving the properties of agricultural plants and animals, although raising important questions about safety. We may also move on to engineering our own bodies, for improved health and welfare.

What about unforeseeable advances? For this, we might look to the gray areas of biology and speculate on areas that might lead to advances as further secrets of nature are revealed.

Thus far, biology appears to act at a level larger than the quantum scale, using deterministic processes. The indeterminacy of quantum mechanics is intimately involved in covalent bonding, reaction kinetics, and electron transfer, but in very predictable ways. We can express these in terms of bond lengths and rates of reaction or transfer, so that none of the miracles of quantum mechanics—such as information traveling faster than the speed of light when quantum mechanical states collapse—need to be taken into account. This does not mean, however, that they cannot be harnessed. Given that bionanomachines operate so close to the quantum scale, they are the perfect candidates for creation of a new quantum technology. Exciting concepts in quantum computing and quantum communication are being studied in theory and in physics laboratories. Bionanomachines may provide the pathway to translate these ideas into practical applications.

Consciousness is still a mystery that may hold unforeseeable surprises in the future. Consciousness is thought by some to be a consequence of complexity, something that will simply appear as soon as our own creations get complicated enough. Others see consciousness as an irreducible property, perhaps a consequence of quantum indeterminacy or perhaps relying on something more metaphysical. As the study of neurobiology expands, thought and memory appear to be solidly rooted in cellular and molecular structure. If consciousness also turns out to be reducible to physical principles, creation of consciousness in artificial objects (beings?) will create diverse opportunities. The subtlety and range of response in biological systems may be the most successful way to create this mysterious property.

LESSONS FOR MOLECULAR NANOTECHNOLOGY

The speculative nanotechnology proposed by Drexler and other molecular nanotechnologists is based on a method of mechanically adding atoms one at a time to a growing structure, through the use of an assembler. Richard E. Smalley has presented two problems with this approach, which he terms the "thick fingers problem" and the "sticky fingers problem."

The "thick fingers problem" is based on the atomicity of all nanomachinery. Because the machine is made of atoms, it cannot have structural details that are finer than the size of the component atoms. This reduces the options for the small constellation of atoms in an assembler that directly interact with the atoms being added to a growing product. Smalley points out that to generate a general-purpose assembler, a variety of chemical environments will be needed and it will prove impossible to design stable robot fingers that can be placed together to create an atom-sized space.

The "sticky fingers problem" is based on the interaction of atoms, which is far different from the interactions of familiar objects. Atoms are sticky. When they get close to one another, they form stable interactions through dispersion forces, hydrogen bonds, and electrostatics. Smalley points out that the interaction with the robot may be just as strong as the interaction with the product, so atoms may remain stuck to the robot arm. Imagine an analogy in our real world. Take a bag of marbles and coat each with a thick layer of rubber cement. Now, with fingers also coated in glue, try to build a pyramid on the table. Challenging.

Of course, the fact that you are sitting here reading these words is proof that both of these problems have solutions. These problems may prove insurmountable for the diamondoid-based mechanosynthesis as envisioned by molecular nanotechnology, but real, working solutions for creating objects from the bottom up by direct manipulation of individual atoms have been developed by biological systems. The solution discovered by nature is hierarchical, using atom-level nanotechnology when possible and self-assembly when it is not.

Enzymes must solve both of these problems. They must create a molecule-sized active site using only atoms, and they must capture their substrates and release their products. Active sites are created with a large overhead of protein infrastructure around the active site, using hundreds of times as many atoms as are needed to interact directly with the substrate. This large infrastructure allows for the precise placement of atoms to form the active site, to tolerances much smaller than the radius of the atom. Products are released by carefully tailoring the binding strength: The active site is designed to bind tightly to the unstable transition state but not to the products. The shape of the active site favors reaction and then release. In many cases, however, the release of products remains the slowest step of the reaction, showing that nature is still plagued by sticky fingers.

Enzymatic assembly is useful for a certain class of reactions. In general, the enzyme must be able to surround the molecules being modified, creating an enclosed chemical environment. This is excellent for creation of custom organic molecules and for creation of linear polymers. But for the creation of large, three-dimensional objects, the approach fails. Enzymes are not generally effective when faced with a flat wall and asked to make changes. So self-assembly is used to create larger structures. Design of protein and nucleic acid polymers that spontaneously fold into stable, globular structure allows the design of modular units, which then self-assemble into objects of any desired size. These 10- to 100-nm modules are far easier to manipulate than individual atoms, and a variety of modification machinery can lay bricks, modify surfaces, activate modules on site, cross-link modules once assembled, and countless other variations.

As we work to design synthetic methods for nanotechnology, there are two key lessons to be learned from biology, lessons that are also learned from chemistry. First, combination of specific atoms into molecules is a difficult and challenging task. In both biology and chemistry, each new molecule, each new bond, requires the design of a custom technique. By all expectations, if we desire to build objects atom by atom, we will have to employ a large set of construction tools tailored specifically for each new assembly task. If, however, we are willing to step up one level coarser and use polymers to build our nanoscale objects, the construction task becomes immeasurably easier. Then a single synthetic reaction may be used in all cases, but a wide variety of monomer units can be used to create a variety of final products. With polymers, we cannot choose any arbitrary combination of atoms. We are limited to the polymeric linkage scheme that we choose. But we gain incredible ease of synthesis and flexibility of design specification. Looking to nature and chemistry, we see a combination of the two techniques: designing specific molecules atom by atom through laborious design of appropriate enzymes and use of proteins, plastics, and other polymers when larger structures are needed.

THREE CASE STUDIES

Natural biomachinery provides working examples of nanomachines in action. Study of molecular biology is an excellent way to get a feeling for how machines work at the nanoscale. The principles are different from anything employed in macroscopic engineering, so our intuition will often play us false. Nanoscale flywheels or eggbeaters that homogenize individual cells are questionable constructs, just as Brownian ratchets are difficult to conceive at the macroscale. By looking to biology, we can see how different functional goals are realized in nanoscale machinery.

In this section, I explore three speculative case studies. These start with a particular nanoscale task that we would like to perform and then look to biology for insight into how to proceed.

Case Study: Nanotube Synthase

Consider a reasonable short-term goal: the creation of a mechanism for synthesizing carbon nanotubes of defined composition. This is a good example of a task at which natural biomachinery could excel. Imagine starting from a simple hydrocarbon ring created by traditional organic chemistry. Then the nanotube synthase adds carbon atoms to the edge, building a nanotube of precise diameter and bonding geometry (Figure 7-1). If the starting rings are immobilized on a substrate, a parallel bundle of nanotubes will be produced. If they are in solution, a random three-dimensional network will result.

Several challenges must be faced. The first is development of a set of reactions for building the nanotube section by section. Many reactions are available for use as models. In particular, many of the carbon addition reactions that use folate as a cofactor might provide models on which to build.

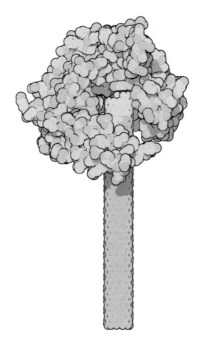

Figure 7-1 This speculative model of a nanotube synthase includes a bound folate molecule, shown in dark pink, that delivers two-carbon units for the reaction. The protein encircles the nanotube, making the reaction more processive and defining the size of the synthetic nanotube.

Folate is a carrier of activated carbon atoms, providing a source of carbon atoms for synthesis and a ready handle for accurate positioning by the enzyme. A folatelike cofactor that carries an ethylene group might provide a ready source of two-carbon stock material for the reaction.

Once the basic reaction is defined, the set of enzymes needed to perform the reaction must be designed. An evolutionary approach, such as catalytic antibodies or SELEX, could be used to search for appropriate molecules or for starting points for further optimization. The tricky part of this approach is the design of an appropriate transition state mimic that is used to select the catalytic antibody or ribozyme. A smaller model system that mimics the reaction, such as the transfer of the two-carbon ethylene group from folate to an aromatic carbon ring, might be used to design the basic enzymatic strategy. This machinery could then be added to the larger synthase complex that both recognizes the leading edge of the tube and performs the synthesis reaction.

The active site must position both the growing nanotube and the new carbon group that will be added. The portion of the nanotube synthase recognizing the stock material is not a problem, because it will most likely be attached to a recognizable carrier such as folate. The portion recognizing the nanotube is more difficult, because the nanotube presents a smooth, carbon-rich surface with no convenient polar atoms for recognition. Examples from fatty acid synthesis and cholesterol synthesis can provide some possible leads. In these cases, the active sites are lined with carbon-rich amino acids, favoring the binding of carbon-rich substrates. The shape is also carefully tailored to fit only to molecules of the desired size. With the nanotube synthase, we have two surfaces to recognize: the outer surface and the inner surface of the tube. By creating a nanotube synthase that encircles the nanotube and perhaps inserts a short loop into the lumen of the tube, the diameter of the tube may be controlled. As an added benefit, the intimate interaction of the enzyme and nanotube may make the enzyme more processive, performing many steps before dissociating and finding a new nanotube.

This goal has the potential to be solved in the next decade or two. All of the necessary tools are becoming available. Evolutionary methods such as catalytic antibodies provide a mechanism for finding the active site machinery, and prediction methods will allow the mechanism to be refined and perfected. Our growing ability to predict protein folding will allow optimization of the design for stability and specificity. And the powerful mechanisms of biotechnology will allow synthesis of large quantities of the final product. Custom enzymes with exotic activities are, by all expectations, just around the corner.

Case Study: A General Nanoscale Assembler

The construction of a molecular assembler is seen by many as being the gateway to nanotechnology. This certainly was the case in nature—once the ribosome was perfected, the evolution of life exploded into the diverse species observed today. The ribosome is a working example of a specific assembler. It creates linear chains of a specific set of building blocks based on a linear information storage medium. What are some considerations for extending nanoscale assembly to more general approaches?

The first need is an *information storage medium.* The ribosome demonstrates the effectiveness of a linear storage medium composed of a small number of atoms. This information may be read directly and used to specify the location of parts and to control the process of construction. It is read like a punched paper tape, specifying a sequential list of instructions. It has the added advantage of acting as a template in some cases, with the actual atomic structure of the storage medium being used to orient the construction apparatus or the modules being connected.

The second need is *positional control.* There is no precedent in nature for general positional control over subnanoscale lengths, of the kind proposed in molecular nanotechnology. The examples from biology would seem to warrant caution about the idea of mechanosynthesis, which requires absolute positioning of about 0.01 nm over a range of 1–10 nm in three dimensions. Biological systems show that positional control may be very fine, but in specific orientations, not over a large range. In bionanomachines, positions are specified relative to template molecules, such as DNA, actin filaments, or enzyme active sites. Enzymes act as defined templates for given reactions and require a large overhead of infrastructure to form the proper molecular geometry to create this template. When motion is needed, it is performed group by group, as in polymerases moving along DNA, or sub-unit by subunit, as in myosin moving along actin.

The success of molecular nanotechnology is contingent on the construction of a molecular assembler with full three-dimensional addressability. With this assembler, individual atoms or molecular fragments are placed in three-dimensional space, building an object of fully defined three-dimensional structure. The only existing information-driven assembler, the ribosome, is a linear assembler, not a three-dimensional assembler. The three-dimensional structure of the products is adopted by folding after release of the linear chain. This is far more limiting than the speculative assemblers, which expand the range of products from self-assembling linear polymers into fully defined three-dimensional constructs.

We might imagine a biologically inspired assembler that uses techniques similar to those of the ribosome to build a three-dimensional lattice product instead of a linear product. The same technique of aligning a specific carrier molecule (the tRNA) is based on instructions in a linear or other storage medium (the mRNA). The assembler would be far more complex, requiring a mechanism for translocating from a given position to another position based on the blueprint. In the case of the ribosome, the translocation is a simple displacement of the last carrier molecule, pushing the RNA one step forward and the growing product polymer one step forward. A similar displacement might be imagined in two or three dimensions, perhaps using carrier molecules that act simply as displacement effectors, which are followed by carriers that add the next subunit once the proper position is adopted.

Of course, there are dozens of potential conceptual problems that must be addressed. Because the ribosome surrounds the informational molecule and guides the product through a narrow tunnel, it is highly processive. It is difficult to imagine a similar geometry that will make a three-dimensional assembler as processive. Also, the ribosome creates a product with no gaps. Most three-dimensional products, however, are not solid blocks. The mechanisms of translocating the assembler from one site to the next must therefore include methods to deal with holes and edges.

The final need is *molecular synthesis*. Of all of the needs for creation of a nanoscale assembler, the chemistry presents the least problem. This problem has been solved, and there are thousands of biological examples of how to perform specific chemical reactions at a given position and at the desired time. Chemical groups may be attached to specific carrier groups to ensure directional transfer of the group. Active site templates may be constructed around the site of attachment, speeding the reaction and reducing unwanted competing reactions. The principles used in natural enzymes may be applied to grip the carrier group, to construct the proper catalytic environment, and to provide the proper environment to discourage the reverse reaction.

Many enzymes take advantage of a modular active site. For instance, nucleotide-binding folds may be combined with other functional sites to add ATP binding to the active site. A similar modularity might be envisioned in a speculative assembler, allowing different functionalities to be swapped for a given synthetic task. The important lesson is that each functional site is composed of a collection of surrounding atoms, creating the proper environment for binding and catalysis. One cannot think of atoms displayed on a functional tip of an assembler; rather, one must think of the assembler as creating the proper environment for positioning and transfer of the atom.

Alternatively, all of the necessary technology is available today for a hybrid approach to a general assembler. The information storage and positional control could be performed with macroscale technology: the atomic force microscope. An enzyme could be added to the tip, where it would perform a nanoscale synthesis on the growing product. We might imagine this assembler working in two modes. It might work in an additive mode, where atoms or molecules are added to a growing product one by one. The enzyme is positioned in the proper place, and the raw materials enter the active site by diffusion from the surrounding medium. We might also use the assembler in a subtractive mode, starting with a solid substrate—crystals, collagen fibers, cellulose, or other solid materials—and using an AFM-positioned digestive enzyme to etch the structure into the substrate. In either case, the process would be laborious, and methods to ensure that each synthetic step has occurred would be necessary.

Case Study: Nanosurveillance

One of the great promises from visionaries in nanotechnology is increased longevity or even immortality. They envision nanorobots that circulate through the body, repairing damage and performing daily maintenance. By remodeling the structures that connect cells, the effects of aging could be countered. By going inside cells and correcting changes, degenerative diseases and cancer could be battled at their source. Amyloid deposits could be cleared, removing the danger of Alzheimer disease. Proteins in the eye lens, one of the few proteins that are built to last for decades, could be repaired, removing the danger of cataracts.

As a case study, let's explore the effort that would be needed to correct one important problem: mutation of the p53 gene that contributes to many forms of cancer. The p53 gene is present in every cell, where it acts as a watchdog to control growth. Mutations in this gene can compromise its function, allowing uncontrolled growth that can lead to cancer. So one duty of a speculative nanorobot would be to watch for mutations in the p53 gene and correct them when they appear.

Several considerations must be addressed. First is the issue of speed. Every cell in the body must be monitored. It would probably be sufficient to check each cell once a year—the effects of p53 mutation take some time to manifest themselves in the form of cancer. The nanorobot will need to enter a cell, find its way to the nucleus, and then examine the entire genome, looking for the p53 gene. For the purposes of this study, assume that the searching, reading, and verifying operations will proceed at speeds similar to the process of DNA replication, about 500 nucleotides per second. At this speed, it would take about 2 months for the nanorobot to scan the entire human genome and to find the p53 gene. Allowing time to travel from cell to cell, this requires about a trillion nanorobots to monitor all of cells in a human being once per year. A trillion nanorobots sounds like a lot, but as Feynman reminded us, molecules (and nanorobots) are really small. If we assume that our nanorobot is about the same size as a ribosome, we will have to infuse the patient with only about 25 µg of nanorobots to allow monitoring each cell once per year. Note that this is a worst-case scenario—if we assume that the nanorobot can find the p53 gene with the same speed as a repressor finds its operator, without unwinding the DNA double helix, it will need only minutes to find its target once inside the nucleus.

The second consideration is information storage. The p53 gene contains about 1500 nucleotides, along with other regions that control synthesis of the protein. All must be verified and/or corrected. As a best-case scenario, we assume that the nanorobot can store this information in a single atom per base; we find that the information storage alone requires a space of about half the size the protein that is being monitored. As a more realistic estimate, if we use a strand of nucleic acid to encode the information, a molecule the same size as the gene must be carried along with the nanorobot to carry the information. For p53, this is equal, roughly, to the size of the small subunit of the ribosome.

Along with the information storage, we must add mechanisms for repair. We will have to replace any mutated bases. We might imagine a simple set of enzymes that excise the faulty base (natural examples of this functionality are available) and then replace it with the proper base. This might involve two or three separate enzyme activities—an excision step, a replacement step, and perhaps an activation step to prime the strand for addition. If a templated approach can be taken, as with the ribosome, the same set of enzymes could be used for all four bases.

The final consideration is entry and exit. Entry into the cell might be accomplished through the normal endocytosis processes of cells, assuming that the nanorobot can withstand the destructive conditions found inside lysosomes. This is the approach taken by many intracellular parasites, including viruses and bacteria, so the keys for entry are available from them. We might design a lipid-enclosed nanorobot that sheds its lipid container after entering the cell. We must be careful, however, because this method of entry often shows differences between cell types, so the resultant nanorobot may be specific for a given type of cell. In some cases, this may be an advantage—in the case of general surveillance of p53, it is not.

Once in the cytoplasm, however, there are no precedents for entry of such a large structure into the nucleus. The nucleus is surrounded by a double membrane pierced by nuclear pores, but these pores are designed for specific transport of protein-sized objects. Our nanorobot will presumably be too large, unless it is designed like a string of beads (this is how large messenger RNA molecules leave the nucleus). In dividing cells, the nuclear membrane is disassembled when the genome is separated. Then the DNA would be directly accessible. However, this would limit the range of the nanorobot to cells that divide, such as skin cells and blood stem cells, missing other classes, such as nerve cells, entirely.

We might look to viruses for methods to leave the cell. Many viruses form buds from the surface of cells. They create a coating of proteins on their own surface that interacts with the inner side of the cell membrane, wrapping the membrane around themselves and pulling themselves out. The bud pinches off, and the virus is left free in the solution surrounded by a mini-membrane. This may be ideal for a nanorobot, because it makes it ready to entry into the next cell.

This analysis reveals why the "repair" paradigm is rarely used in your body. The investment in resources and information is simply too large. Instead, nearly all damage is corrected by replacement. In most cases, proteins have life spans of minutes to weeks, so damaged proteins are quickly degraded and replaced. Damaged cells are destroyed and replaced by entirely new cells from the surrounding areas. A major exception is DNA, which carries information that cannot easily be replaced. A variety of corrective measures are used, but all rely on the damage being local and repairable without reference to another storehouse of information. Systems are available for replacing a nucleotide that is chemically altered, using the complementary strand as the source of information for which base to add.

Creation of a general repair machine designed to maintain, for instance, the entire genome of a cell appears to be infeasible, based on the amount of information that must be carried. That being said, the use of targeted repair machinery may provide extreme benefits. Imagine a set of repair machinery that corrects a few of the major mutations involved in cancer. These might be infused into patients at yearly intervals, performing their repairs and then gracefully breaking down into their component parts within the cells they repair. This type of one-shot nanorobot would also solve one of the major problems with current HIV therapy. HIV integrates into the genome of infected cells and lies dormant for years. No current therapies can find and destroy these latently infected cells, so infection recurs for many years, even in the face of continual drug therapy. A patient could be cleared of infection with a nanorobot that seeks out this integrated DNA and kills any cell that contains it. Ironically, based on the speculation in this case study, this anti-HIV nanorobot might itself look a lot like a virus.

ETHICAL CONSIDERATIONS

Bionanotechnology carries with it a grave responsibility. As with any technology, the potential for misuse is enormous. As articulated by John C. Polkinghorne, bionanotechnology must retain "respect for life and the need for a balance of benefit over harm resulting from any intervention."

We have seen in the past several decades an explosion of technology at all levels—machinery, electronics, information, and biology. Many people have reservations about this fast pace. Some are discouraged by the compulsion toward novelty. Many scientists and engineers explore new technologies simply because they are possible, without spending the time to think about the implications and consequences. Also, many new technologies are the domains of experts and large corporations, which may be pursuing developments for personal motives that do not reflect goals that best benefit humanity or the global environment. The governments of many countries are becoming increasingly aware of the reservations of their populace and are enacting regulations to control the more controversial applications, such as human cloning and embryonic stem cell research. But, of course, it is difficult to decide where to draw the line between morally acceptable technology and applications that are morally reprehensible. As we decide where to draw our own personal line, we might think carefully about two topics: the respect for life and possible dangers.

Respect for Life

Ponder for a moment the incredible hubris of the entire endeavor of bionanotechnology. The natural environment has taken billions of years to perfect the machinery running our bodies and the bodies of all other living things. And in a single generation, we usurp this knowledge and press it to our own use. Today, there is a healthy hesitancy curbing a vast excitement in many scientists and bioengineers. As articulated by Charles Cantor, "The more I reflect on it, the more I am forced to conclude that the next step in humankind's evolution is our acquisition of the power to control the evolution of our own species and all others on this planet. I can only hope that we use this power wisely."

The potential for modifying living organisms, and human beings in particular, provides immediate moral problems. The genetic engineering of children, particularly for cosmetic reasons or to improve native ability, raises severe problems for most people, using the argument that children are not commodities to be picked and sorted through on the department store shelf. However, the ability to remove hereditary diseases, permanently and for all successive generations, has an undeniable appeal.

Of course, this dilemma is not, at its heart, anything new. For centuries, agriculture and medicine have modified biology in profound ways. By selective breeding, we have changed livestock, grains, flowers, dogs, cats, and countless other organisms into grossly different shapes to provide more food and to please our senses. To our own bodies, we add substances to change blood pressure, to fight microorganisms, to relieve pain, and thus extend our life span by decades. Perhaps, in a few decades, the advances of nanotechnology will feel as familiar as a hybrid tea rose.

Potential Dangers

The potential dangers of nanotechnology are a favorite topic in current science fiction. In particular, the concept of the rogue disassembler/assembler has been widely discussed, both in fiction and by speculative scientists. We have abundant precedents for how to proceed (and warnings of how not to proceed) from other technologies that pose dangers when used improperly. These include regulations on research in nuclear science and viral research that may be applied to sensitive applications in bionanotechnology.

Addressing potential dangers can lead to additional moral complications. Take, for instance, the incorporation of terminator genes into genetically modified seeds that make them sterile and productive only for a single generation. Although this provides a ready solution to the possible spread of the engineered plant, it has been criticized as a method to ensure continued sales, as farmers will require new seed for each year's crop. This provides a significant hardship for farmers in developing countries, where seed is typically saved from one season to the next, despite the fact that this is the market often advertised as the major winners for these modified crops.

On a more familiar level, one is faced with the question of the need for intervention. Just because we have a technology, we are not obligated to use it. Modern pharmaceuticals are extending their range from the traditional realm of infectious disease and life-threatening disorders into cosmetic and lifestyle changes. One might question the need for medications for weight control, stress, and even jet lag when behavioral modification can curb the negative effects in most cases. Medications for reducing the signs of aging and improving sexual performance can add significantly to quality of life, but at the cost of side effects. In our present society, individuals are given the choice for access to these elective treatments, although often without the proper amount of information on likely consequences. As bionanotechnology proceeds and ever more powerful techniques are developed for modifying human biological function, the need for intervention will become an even greater question and will require careful education of users of the technology.

Final Thoughts

The potential of bionanotechnology for feeding the world, for improving our health, for providing rapid and cheap manufacturing with environmental mindfulness, is immense. But we must temper this excitement with careful thought. The philosophy of manifest destiny, that all things are provided for our use, is an integral part of Western culture, and is often followed without introspection and thought as to the potential outcomes. It would serve us well to look to Nature—to her world-spanning interconnectedness, to her unassuming creativity, to the sheer wonder of her accomplishments—for guidance as we proceed, tempering the strong cultural forces of novelty and capital gain.

LITERATURE

GENERAL

Alberts, B., Johnson, A., Lewis, J., Raff, M., Roberts, K. and Walter, P. (2002) *Molecular Biology of the Cell*. Garland Science, New York.

Drexler, K.E. (1992) *Nanosystems: Molecular Machinery, Manufacturing, and Computation*. John Wiley & Sons, New York.

Goodsell, D.S. (1993) *The Machinery of Life*. Springer, New York.

Goodsell, D.S. (1996) *Our Molecular Nature: The Body's Motors, Machines and Messages*. Springer, New York.

Goodsell, D.S. (2000) "Biomolecules and Nanotechnology." *American Scientist* 88, 230–237.

Stryer, L. (1995) *Biochemistry*. W.H. Freeman, New York.

CHAPTER 2

The Unfamiliar World of Biomolecules

Hess, B. and Mikhailov, A. (1995) "Microscopic Self-organization in Living Cells: A Study of Time Matching." *Journal of Theoretical Biology* 176, 181–184.

Purcell, E.M. (1997) "Life at Low Reynolds Number." *American Journal of Physics* 45, 3–11.

Biomolecules

Branden, C. and Tooze, J. (1991) *Introduction to Protein Structure*. Garland, New York.

Creighton, T.E. (1993) *Proteins: Structures and Molecular Properties*. W. H. Freeman, New York.

Dennis, C., editor. (2003) "The Double Helix—50 Years." *Nature* 421, 395–453.

Dickerson, R.E. and Geis, I. (1969) *The Structure and Action of Proteins*. Harper and Row, New York.

Lipowsky, R. and Sackmann, E., editors. (1995) *Handbook of Biological Physics. Volume 1: Structure and Dynamics of Membranes*. Elsevier Science, Amsterdam.

Bionanotechnology: Lessons from Nature. By David S. Goodsell
ISBN 0-471-41719-X   Copyright © 2004 John Wiley & Sons, Inc.

CHAPTER 3

Recombinant DNA Technology

Ausubel, F.M., Brent, R., Kingston, R.E., Moore, D.D., Seidman, J.G., Smith, J.A. and Struhl, K., editors. (1999) *Short Protocols in Molecular Biology*, 4th Ed. Wiley, New York.

Jermutus, L., Lyubov, L.A. and Pluckthun, A. (1998) "Recent Advances in Producing and Selecting Functional Proteins by Using Cell-Free Translation." *Current Opinion in Biotechnology* 9, 534–548.

Rudolph, R. (1996) "Successful Protein Folding on an Industrial Scale." In *Protein Engineering: Principles and Practice*, J.L. Cleland and C.S. Craik, eds. Wiley-Liss, New York.

Structure Determination

Baumeister, W., Grimm, R. and Walz, J. (1999) "Electron Tomography of Molecules and Cells." *Trends in Cell Biology* 9, 81–85.

Blundell, T.L. and Johnson, L.N. (1976) *Protein Crystallography*. Academic, New York.

Bustamante, C. and Keller, D. (1995) "Scanning Force Microscopy in Biology." *Physics Today*, December 1995, 32–38.

Clore, G.M. and Gronenborn, G.M. (1998) "NMR Structure Determination of Proteins and Protein Complexes Larger than 20 kD." *Current Opinion in Chemical Biology* 2, 564–570.

Nogales, E. and Grigorieff, N. (2001) "Molecular Machines: Putting the Pieces Together." *Journal of Cell Biology* 152, F1–F10.

Protein Structure Prediction

Rost, B. and Sander, C. (1996) "Bridging the Sequence-Structure Gap by Structure Predictions." *Annual Review of Biophysics and Biomolecular Structure* 25, 113–136.

Molecular Modeling

Taylor, R.D., Jewsbury, P.J. and Essex, J.W. (2002) "A Review of Protein-Small Molecule Docking Methods." *Journal of Computer-Aided Molecular Design* 16, 151–166.

CHAPTER 4

Biomolecular Structure and Stability

Eisenberg, D. and Crothers, D. (1979) *Physical Chemistry with Applications to the Life Sciences*. Benjamin/Cummings, Menlo Park, CA.

Protein Folding

DeGrado, W.F., Summa, C.M., Pavone, V., Nastri, F. and Lombardi, A. (1999) "De Novo Design and Structural Characterization of Proteins and Metalloproteins." *Annual Review of Biochemistry* 68, 779–819.

Richardson, J.S. (1981) "The Anatomy and Taxonomy of Protein Structure." *Advances in Protein Chemistry* 34, 167–339.

Vieille, C. and Zeikus, G.J. (2001) "Hyperthermophilic Enzymes: Sources, Uses and

Molecular Mechanisms for Thermostability." *Microbiology and Molecular Biology Reviews* 65, 1–43.

Self-Assembly
Caspar, D.L.D. and Klug, A. (1962) "Physical Principles in the Construction of Regular Viruses." *Cold Spring Harbor Symposia on Quantitative Biology* 27, 1–24.

Ellis, R.J. (2001) "Macromolecular Crowding: An Important but Neglected Aspect of the Intracellular Environment." *Current Opinion in Structural Biology* 11, 114–119.

Goodsell, D.S. and Olson, A.J. (2000) "Structural Symmetry and Protein Function." *Annual Review of Biophysics and Biomolecular Structure* 29, 105–153.

Johnson, J.E. and Speir, J.A. (1997) "Quasi-equivalent Viruses: A Paradigm for Protein Assemblies." *Journal of Molecular Biology* 269, 665–675.

Whitesides, G.M., Mathias, J.P. and Seto, C.T. (1991) "Molecular Self-Assembly and Nanochemistry: A Chemical Strategy for the Synthesis of Nanostructures." *Science* 254, 1312–1319.

Self-Organization
Lipowsky, R. and Sackmann, E., editors. (1995) *Handbook of Biological Physics. Volume 1: Structure and Dynamics of Membranes.* Elsevier Science, Amsterdam.

Molecular Recognition
Crane, H.R. (1950) "Principles and Problems of Biological Growth." *The Scientific Monthly* 70, 376–389.

Larsen, T.A., Olson, A.J. and Goodsell, D.S. (1998) "Morphology of Protein-Protein Interfaces." *Structure* 6, 421–427.

Zhao S., Morris, G.M., Olson, A.J. and Goodsell, D.S. (2001) "Recognition Templates for Predicting Adenylate-binding Sites in Proteins." *Journal of Molecular Biology* 314, 1245–1255.

Flexibility
Karplus, M. and McCammon, J.A. (1983) "Dynamics of Proteins: Elements and Function." *Annual Review of Biochemistry* 53, 263–300.

Wright, P.E. and Dyson, J.E. (1999) "Intrinsically Unstructured Proteins: Re-assessing the Protein Structure-Function Paradigm." *Journal of Molecular Biology* 293, 321–331.

CHAPTER 5

Information-Driven Nanoassembly
Cox, J.P.L. (2001) "Long-Term Data Storage in DNA." *Trends in Biotechnology* 19, 247–250.

Puglisi, J.D., Blanchard, S.C. and Green, R. (2000) "Approaching Translation at Atomic Resolution." *Nature Structural Biology* 7, 855–861.

Energetics
Fink, H.-W. (2001) "DNA and Conducting Electrons." *Cellular and Molecular Life Sciences* 58, 1–3.

Page, C.C., Moser, C.C., Chen, X. and Dutton, P.L. (1999) "Natural Engineering Principles of Electron Tunnelling in Biological Oxidation-Reduction." *Nature* 402, 47–52.

Porath, D., Bezryadin, A., deVries, S. and Dekker, C. (2000) "Direct Measurement of Electrical Conduction through DNA Molecules." *Nature* 403, 635–638.

Chemical Transformation

Sinnott, M., editor. (1998) *Comprehensive Biological Catalysis: A Mechanistic Reference.* Academic, San Diego, CA.

Regulation

Monod, J., Wyman, J. and Changeaux, J.-P. (1965) "On the Nature of Allosteric Transitions: A Plausible Model." *Journal of Molecular Biology* 12, 88–118.

Perutz, M.F. (1989) "Mechanisms of Cooperativity and Allosteric Regulation in Proteins." *Quarterly Reviews of Biophysics* 22, 139–236.

Biomaterials

Addadi, L. and Weiner, S. (1992) "Control and Design Principles in Biological Mineralization." *Angewandte Chemie* 31, 153–169.

Petka, W.A., Harden, J.L., McGrath, K.P., Wirtz, D. and Tirrell, D.A. (1998) "Reversible Hydrogels from Self-Assembling Artificial Proteins." *Science* 281, 389–392.

Rief, M., Gautel, M., Oesterhelt, F., Fernandez, J.M. and Gaub, H.E. (1997) "Reversible Unfolding of Individual Titin Immunoglobulin Domains by AFM." *Science* 276, 1109–1112.

Tskhovrebova, L., Trinick, J., Sleep, J.A. and Simmons, R. M. (1997) "Elasticity and Unfolding of Single Molecules of the Giant Muscle Protein Titin." *Nature* 387, 308–312.

Waite, J.H. (1987) "Nature's Underwater Adhesive Specialist." *International Journal of Adhesion and Adhesives* 7, 9–14.

Motors

Bray, D. (2001) *Cell Movements: From Molecules to Motility.* Garland, New York.

Bustamante, C., Keller, D. and Oster, G. (2001) "The Physics of Molecular Motors." *Accounts of Chemical Research* 34, 412–420.

DeRosier, D.J. (1998) "The Turn of the Screw: The Bacterial Flagellar Motor." *Cell* 93, 17–20.

Oster, G. and Wang, H. (1999) "ATP Synthase: Two Motors, Two Fuels." *Structure* 7, R67–R72.

Pantaloni, D., LeClainche, C. and Carlier, M.F. (2001) "Mechanism of Actin-based Motility." *Science* 292, 1502–1506.

Squire, J.M. (1997) "Architecture and Function in the Muscle Sarcomere." *Current Opinion in Structural Biology* 7, 247–257.

Vale, R.D. and Milligan, R.A. (2000) "The Way Things Move: Looking Under the Hood of Molecular Motor Proteins." *Science* 288, 88–96.

Traffic Across Membranes

Doyle, D.A., Cabral, J.M., Pfuetzner, R.A., Kuo, A., Gulbis, J.M., Cohen, S.L., Chait,

B.T. and MacKinnon, R. (1998) "The Structure of the Potassium Channel: Molecular Basis of K+ Conduction and Selectivity." *Science* 280, 69–77.

Kuhlbrandt, W. (2000) "Bacteriorhodopsin, the Movie." *Nature* 406, 569–570.

Biomolecular Sensing

Anson, L., editor. (2001) "Molecular Sensing" *Nature* 413, 185–230.

Batiza, A.F., Rayment, I. and Kung, C. (1999) "Channel Gate! Tension, Leak and Disclosure." *Structure* 7, R99–R103.

Self-Replication

Fraser, C.M. et al. (1995) "The Minimal Gene Complement of *Mycoplasma genitalium*." Science 270, 397–403.

CHAPTER 6

Basic Capabilities

Dahiyat, B.I. and Mayo, S.L. (1997) "De novo Protein Design; Fully Automated Sequence Selection." *Science* 278, 82–87.

DeGrado, W.F., Summa, C.M., Pavone, V., Nastri, F. and Lombardi, A. (1999) "De Novo Design and Structural Characterization of Proteins and Metalloproteins." *Annual Review of Biochemistry* 68, 779–819.

Gilmore, M.A., Steward, L.E. and Chamberlin, A.R. (1999) "Incorporation of Noncoded Amino Acids by In Vitro Protein Biosynthesis." *Topics in Current Chemistry* 202, 77–99.

Gutte, B. (1977) "Study of RNase A Mechanism and Folding by Means of Synthetic 63-Residue Analogs." *Journal of Biological Chemistry* 252, 663–670.

Richardson, J.S. and Richardson, D.C. (1989) "The de novo Design of Protein Structures." *Trends in Biochemical Sciences* 14, 304–309.

Riddle, D.S., Santiago, J.V., Bray-Hall, S.T., Doshi, N., Grantsharova, V.P., Yi, Q. and Baker, D. (1997) "Functional Rapidly Folding Proteins From Simplified Amino Acid Sequences." *Nature Structural Biology* 4, 805–809.

Uhlmann, E., Peyman, A., Breipohl, G. and Will, D.W. (1998) "PNA: Synthetic Polyamide Nucleic Acids with Unusual Binding Properties." *Angewandte Chemie* 37, 2796–2823.

Wang, L., Brock, A., Herberich, B. and Schultz, P.G. (2001) "Expanding the Genetic Code of *Escherichia coli*." *Science* 292, 498–500.

Harnessing Evolution

Gold L., Polinsky B., Uhlenbeck, O. and Yarius M. (1995) "Diversity of Oligonucleotide Functions." *Annual Review of Biochemistry* 64, 763–797.

Hilvert, D. (1999) "Stereoselective Reactions with Catalytic Antibodies." *Topics in Stereochemistry* 22, 83–135.

Schultz, P.G. and Lerner, R.A. (1995) "From Molecular Diversity to Catalysis: Lessons from the Immune System." *Science* 269, 1835–1842.

Soukup, G.A. and Breaker, R.R. (2000) "Allosteric Nucleic Acid Catalysts." *Current Opinion in Structural Biology* 10, 318–325.

Whaley, S.R., English, D.S., Hu, E.L., Barbary, P.F. and Belcher, A.M. (2000)

"Selection of Peptides with Semiconductor Binding Specificity for Directed Nanocrystal Assembly." *Nature* 405, 665–668.

Wilson, D.S. and Szostak, J.W. (1999) "In vitro Selection of Functional Nucleic Acids." *Annual Review of Biochemistry* 68, 611–647.

Self-Assembly

Ball, P. (1994) *Designing the Molecular World*. Princeton University Press, Princeton, NJ.

Bowden, N.B., Weck, M., Choi, I.S. and Whitesides, G.M. (2001) "Molecule-Mimetic Chemistry and Mesoscale Self-Assembly." *Accounts of Chemical Research* 34, 231–238.

Conn, M.M. and Rebek, J. (1997) "Self-Assembling Capsules." *Chemical Reviews* 97, 1647–1668.

Granja, J.R. and Ghadiri, M.R. (1994) "Channel-Mediated Transport of Glucose Across Lipid Bilayers." *Journal of the American Chemical Society* 116, 10785–10786.

Liu, F., Sha, R. and Seeman, N.C. (1999) "Modifying the Surface Features of Two-Dimensional DNA Crystals." *Journal of the American Chemical Society* 121, 917–922.

Padilla, J.E., Colovos, C. and Yeates, T.O. (2001) "Nanohedra: Using Symmetry to Design Self-Assembling Protein Cages, Layers, Crystals, and Filaments." *Proceedings of the National Academy of Sciences USA* 98, 2217–2221.

Seeman, N.C. (1998) "DNA Nanotechnology: Novel DNA Constructions." *Annual Review of Biophysical and Biomolecular Structure* 27, 225–248.

Service, R.F., Szuromi, P. and Uppenbrink, J., editors. (2002) "Strength in Numbers, Special Section of Supramolecular Chemistry and Self-Assembly." *Science* 295, 2395–2421.

Molecular Motors

Noji, H., Yasuda, R., Yoshida, M. and Kinosita, K. (1997) "Direct Observation of the Rotation of F1-ATPase." *Nature* 386, 299–302.

Soong, R.K., Bachand, G.D., Neves, H.P., Olkhovets, A.G., Craighead, H.G. and Montemagno, C.D. (2000) "Powering an Inorganic Nanodevice with a Bio-molecular Motor." *Science* 290, 1555–1558.

Yurke, B., Turberfield, A.J., Mills, A.P., Simmel, F.C. and Neumann, J.L. (2000) "A DNA-Fuelled Molecular Machine Made of DNA." *Nature* 406, 605–608.

Yan, H., Zhang, X., Shen, Z. and Seeman, N.C. (2002) "A Robust DNA Mechanical Device Controlled by Hybridization Topology." *Nature* 415, 62–65.

DNA Computation

Adleman, L.M. (1994) "Molecular Computation of Solutions to Combinatorial Problems." *Science* 266, 1021–1024. [See also the letters and response in *Science* 268, 481–484 (1995).]

Benenson, Y., Paz-Elizur, T., Adar, R., Keinan, E., Livneh, Z. and Shapiro, E. (2001) "Programmable and Autonomous Computing Machine Made of Biomolecules." *Nature* 414, 430–434.

Braich, R.S., Chelyapov, N., Johnson, C., Rothemund, P.W.K. and Adleman, L. (2002) "Solution of a 20-Variable 3-SAT Problem on a DNA Computer." *Science* 296, 499–502.

Ruben, A.J. and Landweber, L.F. (2000) "The Past, Present and Future of Molecular Computing." *Nature Reviews: Molecular Cell Biology* 1, 69–72.

Nanomedicine

Chang, T.M.S. (1999) "Future Prospects for Artificial Blood." *Trends in Biotechnology* 17, 61–67.

Friedman, T. (1997) "Overcoming the Obstacles of Gene Therapy." *Scientific American* 276(6), 96–101.

Ginsberg, G.S. and McCarthy, J.J. (2001) "Personalized Medicine: Revolutionizing Drug Discovery and Patient Care." *Trends in Biotechnology* 19, 491–496.

Nielsen, P.E. (1999) "Applications of Peptide Nucleic Acids." *Current Opinion in Biotechnology* 10, 71–75.

Hybrid Materials

Brott, L.L., Naik, R.R., Pikas, D.J., Kirkpatrick, S.M., Tomlin, D.W., Whitlock, P.W., Clarson, S.J. and Stone, M.O. (2001) "Ultrafast Holographics Nanopatterning of Biocatalytically Formed Silica." *Nature* 413, 291–293.

Cha, J.N., Stucky, G.D., Morse, D.E. and Deming, T.J. (2000) "Biomimetic Synthesis of Ordered Silica Structures Mediated by Block Copolymers." *Nature* 403, 289–292.

Keren, K., Krueger, M., Gilad, R., Ben-Yoseph, G., Sivan, U. and Braun, E. (2002) "Sequence-Specific Molecular Lithography on Single DNA Molecules." *Science* 297, 72–75.

Storhoff, J.J. and Mirkin, C.A. (1999) "Programmed Materials Synthesis with DNA." *Chemical Reviews* 99, 1849–1862.

Delivery and Transport

Lasic, D.D. (1995) "Applications of Liposomes." In *Handbook of Biological Physics. Volume 1: Structure and Dynamics of Membranes*. R. Lipowsky and E. Sackmann, eds. Elsevier Science, Amsterdam.

Biosensors

Cass, A.E.G., Davis, G., Francis, G.D., Hill, H.A.O., Aston, W.J., Higgins, I.J., Plotkin, E.V., Scott, L.D.L. and Turner, A.P.F. (1984) "Ferrocene-Mediated Enzyme Electrode for Amperometric Determination of Glucose." *Analytical Chemistry* 56, 667–671.

Chen, T., Barton, S.C., Binyamin, G., Gao, Z., Zhang, Y., Kim, H.-H. and Heller, A. (2001) "A Miniature Biofuel Cell." *Journal of the American Chemical Society* 123, 8630–8631.

Howorka, S., Cheley, S. and Bayley, H. (2001) "Sequence-Specific Detection of Individual DNA Strands Using Engineered Nanopores." *Nature Biotechnology* 19, 636–623.

Kricka, L.J. (2001) "Principles of Immunochemical Techniques." In *Tietz Fundamentals of Clinical Chemistry*, 5th Ed. C.A. Burtis and E.R. Ashwood, eds. W. B. Saunders, Philadelphia, PA.

Artificial Life

Cello, J., Paul, A.V., and Wimmer, E. (2002) "Chemical Synthesis of Poliovirus cDNA: Generation of Infectious Virus in the Absence of Natural Template." *Science* 297, 1016–1018.

Chang, T.M.S. (1985) "Artificial Cells Containing Multienzyme Systems." *Methods in Enzymology* 112, 195–203.

Dickerson, R.E. (1978) "Chemical Evolution and the Origin of Life." *Scientific American* 239(3), 70–86.

Joyce, G. F. (2002) "The Antiquity of RNA-Based Evolution." *Nature* 418, 214–221.

Steinberg-Yfrach, G., Rigaud, J.-L., Durantini, E. N., Moore, A. L., Gust, D. and Moore, T.A. (1998) "Light-Driven Production of ATP Catalyzed by F_0F_1-ATP Synthase in and Artificial Photosynthetic Membrane." *Nature* 392, 479–482.

Walde, P., Wick, R., Fresta, M., Mangone, A. and Luisi, P.L. (1994) "Autopoietic Self-Reproduction of Fatty Acid Vesicles." *Journal of the American Chemical Society* 116, 11649–11654.

CHAPTER 7

Cantor, C.R. (2000) "Biotechnology in the 21st Century." *Trends in Biotechnology* 18, 6–7.

Ethical Issues

Polkinghorne, J.C. (2000) "Ethical Issues in Biotechnology." *Trends in Biotechnology* 18, 8–10.

SOURCES FOR BIOMOLECULAR STRUCTURES

Most structures were taken from the Protein Data Bank at http://www.pdb.org. Four letter accession codes for each structure are given below, and full references to these structures are available in the PDB files.

1-1   ribosome 1fjg, 1jj2

2-1   hemoglobin 1hho
2-2   triose phosphate isomerase 7tim (unfolded manually)
2-3   tropomyosin 1c1g; porin 1prn
2-4   collagen 1bkv; insulin 4ins; porin 1prn; trypsin 2ptc; calmodulin 3cln; GCN4 1ysa; ferredoxin 1fxd
2-5   ideal DNA generated in InsightII
2-8   thymidine synthase 1lce
2-9   RNA polymerase 1i6h
2-10   ribosome 1gix, 1giy
2-11   ATP synthase 1c17, 1e79
2-13   opsin 1f88
2-16   collagen 1a3i

3-1   ecoRI 1eri; DNA ligase 1dgs
3-3   taq polymerase 1tau
3-5   lysozyme 1lyd; mutant lysozyme 1l35
3-6   antibody 1igt; exotoxin A 1ikq
3-7   antibodies 1igt, 1bql, 1emt
3-9   lysozyme 4lyz, 2lzt
3-10   lysozyme 1e8l
3-13   lysozyme 1lyd
3-14   HIV protease 1hxb

4-1   satellite tobacco necrosis virus 2stv

4-2   insulin 4ins

4-8   superoxide dismutase 2sod

4-9   lysozyme 1lz1

4-10   satellite tobacco necrosis virus 2stv

4-11   protein folds 2ccy, 1mbn, 1lrv, 1ppr, 1ccm, 1fbr, 1vie, 1prn, 4bcl, 1stm, 1hcd, 1jpc, 1rie, 1got, 1air, 1ndd, 1tim, 1kvd, 1fua, 2dnj

4-12   prefoldin 1fxk; GroEL-GroES 1aon

4-14   pepsin 5pep; Max protein 1an2; porin 2por; potassium channel 1f6g; phosphofructokinase 1pfk; aspartate carbamoyltransferase 1at1; glycolate oxidase 1gox; glutamine synthetase 2gls; protocatechuate 3,4-dioxygenase 3pcg; ferritin 1hrs; satellite tobacco necrosis virus 2stv

4-16   β-tryptase 1a0l; phosphofructokinase 1pfk

4-17   actin 1atn; microtubule 1tub

4-18   satellite tobacco necrosis virus 2stv; tomato bushy stunt virus 2tbv

4-20   ras protein 121p; cytochrome oxidase 1oco

4-21   enolase 3enl

4-22   cAMP-dependent protein kinase 1atp

4-23   isoleucyl-tRNA synthetase 1qu2

4-24   thermolysin 1hyt, 2tli

4-25   adenylate kinase 1ank, 4ake

5-1   insulin 4ins; sequence from J00265 at NCBI, http://www.ncbi.nlm.nih.gov

5-2   ribosome 1fjg, 1jj2

5-4   ribosome 1gix, 1giy; elongation factors 1dar, 1ttt; aminoacyl-tRNA synthetases 1asy, 1eiy, 1ffy

5-6   aspartyl-tRNA synthetase 1c0a; myosin 1br1

5-7   ferredoxin 1roe; photosystem 1jb0; plastocyanin 1bxu

5-8   photosystem 1jb0

5-9   ferredoxin 1blu; nitrogenase 1n2c; cytochrome b-c_1 complex 1bgy

5-10   cytochrome b-c_1 complex 1bgy; cytochrome c 3cyt; cytochrome oxidase 1oco

5-11   photosystem 1jb0; ferredoxin 1roe; plastocyanin 1bxu; cytochrome b-c_1 complex 1bgy; ATP synthase 1c17, 1e79

5-12   triose phosphate isomerase 2ypi

5-15   hemoglobin oxy 1hho; deoxy 2hhb

5-16   fructose bisphosphatase 4fbp, 5fbp

5-17   src protein 2src; pepsinogen 3psg; pepsin 5pep

5-26   myosin 1br1, 2mys

5-28   kinesin 2kin

5-29   ATP synthase 1c17; 1e79

5-35   porin 1prn

5-36   potassium channel 1k4c

5-37   ABC transporter 1l7v

5-36   bacteriorhodopsin 1b3w, 1dze

5-37   mscL channel, from the supplementary information in Sukharev S., Betanzos M., Chiang C.S. and Guy H.R. (2001) *Nature* 409, 720–724.

6-1   ribonuclease A 1rta

6-2   α3D 2a3d

6-3   pna 1o12
6-4   HIV protease 1hsg, 1hxb, 1hxw, 1ohr
6-6   hemoglobin 1hho
6-9   calixarenes GRJRAE, ADACIR from CCDC, http://www.ccdc.cam.ac.uk
6-12 ATP synthase 1e79; streptavidin 1swe
6-15 catalytic antibody 1c1e
6-16 theophylline aptamer 1o15
6-19 poliovirus 2plv
6-24 hemolysin 7ahl

INDEX

Bionanotechnology: Lessons from Nature. By David S. Goodsell
ISBN 0-471-41719-X   Copyright © 2004 John Wiley & Sons, Inc.